The Wireless Officer

by

Percy F. Westerman

The Wireless Officer
by Percy F. Westerman

Copyright © 2023

All Rights reserved.

ISBN: 978-93-59958-04-0

Published by

DOUBLE 9 BOOKS
2/13-B, Ansari Road
Daryaganj, New Delhi – 110002
info@double9books.com
www.double9books.com
Tel. 011-40042856

ABOUT THE AUTHOR

English writer Percy Francis Westerman wrote a number of kid's books. He changed into born in 1876 and died on February 22, 1959. A lot of his books are action tales with military or naval issues. He changed into born in Portsmouth, England, in 1876 and went to Portsmouth Grammar School. When he changed into twenty, he got a process as a clerk at Portsmouth Dockyard. It became in 1900 that he married Florence Wager of Portsmouth. They cherished crusing so much that they spent some of their holiday crusing within the Solent. John F.C. Westerman, their son who become born in 1901, also wrote boys' journey books. He needed to depart his houseboat for dry land due to a fall while he changed into 70 years' vintage, but he kept writing fast. He died at the age of eighty-two, and his final book, Mistaken Identity, got here out in 1959 after he had died. He is said to have started out writing after he and his spouse bet sixpence that he ought to write a better tale than the only he changed into studying to their ill son on the time. He wrote his first book for boys referred to as A Lad of Grit, which came out in 1908 with Blackie and Son Limited. In the identical yr, Baden-Powell started out the Scouting motion, which had a huge impact on a lot of Westerman's books. He became a large fan of the Sea Scouts mainly.

CONTENTS

CHAPTER I
Good News

"Hurrah!" exclaimed Peter Mostyn. "Now, this *does* look like business."

"It does, Mr. Mostyn," agreed the postmistress. "It seems as if we are to lose you again."

"And about time too," added the youth, as he ripped open the long, buff-coloured envelope bearing the words "Broughborough International Marine Telegraph Company".

Peter Mostyn had been "on the beach" for nearly six months. In other words, he was out of a berth. Not that it was any fault of his that a promising and energetic young wireless officer should be without a ship for such a protracted period. An unprecedented slump in British shipping—when hundreds of vessels flying the Red Ensign were laid up, while the bulk of the world's trade was carried by the mercantile fleet of Germany—had resulted, amongst other ills, in the wholesale "sacking" of officers and men, who to a great extent had been the means of warding off the grim spectre of starvation during those black years of the World War.

Five times a week for over four months Peter Mostyn had ridden into Trentham Regis village in the hope of securing the long-expected missive giving him an appointment to another ship.

And now the anxiously awaited communication had arrived. The postmistress, a good, kindly soul to whom the affairs of every man, woman, and child in the Trentham Regis postal area were open secrets, was almost as excited as the recipient, when she handed the envelope over the counter between the piles of groceries that comprised the commercial side of the shop.

For a brief instant Peter was assailed by the dread that the envelope would contain a stereotyped announcement to the effect that his application was still under consideration; then a look of undisguised relief and gratification overspread his tanned features.

"Yes, Mrs. Young!" he exclaimed; "I'm off in three days' time. Where? I hardly know, but I rather fancy it's East Africa. Good evening."

Leaving the postmistress to spread the news amongst the good folk of Trentham Regis—a feat that she would certainly accomplish to her unbounded satisfaction before the post office closed for the night—Peter mounted his motor-bicycle and rode after the fashion of the long departed Jehu, the son of Nimshi, leaving behind him a long trail of chalky dust in the still evening air. Short of wireless it was doubtful whether the good news could have reached "The Pines" in less time, for within the space of five and a half minutes Peter had covered the three miles that separated his home from Trentham Regis.

"Hello, Mater!" he exclaimed, bursting into the house like a young typhoon. "Where are you? Ah, there you are! I've got it!"

There was no need for Mrs. Mostyn to ask for further enlightenment upon the cryptic "it". She guessed the news at once.

"I'm so glad, Peter!" she exclaimed. "What ship this time?"

"The *West Barbican*, Mater. I fancy she's one of the Blue Crescent Line. If so, it's East Africa and possibly India, this trip. 'Tany rate, I'm to join her before noon on Thursday. Where's the Pater?"

"Not back from town yet. There was a board meeting on this afternoon."

"Oh yes, I remember. About the Kilba Protectorate contract. I suppose he'll return by the 8.50.... By Jove! What a topping stunt! Fancy getting a ship again after all this time. Of course, Mother, it was nice to be home again, but, after all, it's a jolly long time to be kicking one's heels on the beach, isn't it?"

Mrs. Mostyn agreed, but solely upon her son's account. She was in no hurry to send her boy to sea again, but she realized that it was for his good that he should once more adventure upon the ocean. Coming of a seafaring family that for generations past had sent its sons down to the sea in ships— often never to return—she realized more than a good many mothers what was meant by the call of the great waters. She had drunk deeply of the cup of sorrow when the S.S. *Donibristle*, of which Peter was wireless officer, had been reported overdue and missing, and was afterwards given out by Lloyd's as a total loss. And in due course Peter had returned home, modestly making light of his hairbreadth adventures, his chief complaint being that the Broughborough International Marine Telegraph Company had not thought fit to appoint him to a ship belonging to the same fleet as did the S.S. *Donibristle*.

Peter's father, John Mostyn, was a retired Engineer Captain, R.N., who, having severed his connection with the navy at the conclusion of the Great War, had become one of the managing directors of the Brocklington Ironworks Company.

For a while the affairs of the newly formed company had flourished. Then came the inevitable slump. Labour troubles and foreign competition added to the difficulties of the firm. The reserve capital dwindled until there were barely sufficient funds to meet the weekly wages bill. Things looked black—decidedly so; but never once did the gloomy outlook daunt the cheery optimism of retired Engineer Captain John Mostyn.

When the fortunes of the Brocklington Ironworks Company seemed at their lowest ebb, the firm had an invitation to tender for a big contract for the recently formed Kilba Protectorate. Comprising a vast extent of territory on the East Coast of Africa, Kilba was making a bold bid for prosperity under British rule. Amongst other schemes for the development of the country was the proposed construction of a railway linking up the coast with the rich mineral lands of the interior. One of the natural difficulties in the way of the railroad was the Kilembonga Gorge, where the river of that name flows with great impetuosity between sheer walls of rock rising two hundred feet above the turgid stream. The bed of the river was of shifting sand, so that, even if the difficulty of the current could be overcome, there remained the question of how to build solid piers on such a doubtful foundation. Finally it was decided to throw a cantilever bridge across the chasm.

Accordingly, the Kilba Protectorate Government asked tenders for the construction of the necessary steelwork, including delivery upon the site. British, American, Italian, Japanese, and German firms were invited to contract, although it was difficult to see why the Kilba Government should have asked these last to quote a price. It was not until later that a reason was forthcoming.

Among the British firms to tender was the Brocklington Ironworks Company, and it was then that Captain Mostyn seized his opportunity. He foresaw that a successful carrying out of the contract would be the turning-point of the firm's fortunes—that the kudos derived from that prosperous enterprise would give the Brocklington Ironworks Company a world-wide advertisement and place them at the forefront of engineering contracting firms.

Upon putting the circumstances of the case before his brother-directors, Captain Mostyn carried his point. He told them that the immediate financial results of the contract would be small—in fact, almost insignificant—but once having beaten all rivals, British and foreign, the future success of the company was assured.

"Capital?" echoed Captain Mostyn, in answer to a question from one of his fellow-directors. "Capital? We can find the capital. It will be a tight squeeze, a terribly tight squeeze, but we'll do it with a slight margin to spare. Let me have a talk with the men, and I'll warrant that, if they have the good sense I credit them with, we'll pull the thing off successfully."

Without delay the managing director went down to Brocklington, where he had what he called a straight talk with the firm's employees. He told them straight that if their whole-hearted co-operation were not forthcoming the works would have to close down, and that, with the present state of unemployment, it would be difficult, almost impossible, for the six hundred hands to find work elsewhere.

"I'm asking you to make sacrifices," he went on. "For the present neither the directors nor the shareholders are making money, and naturally we cannot run this business as a charity concern. I therefore propose a general reduction of wages in order for us to quote competitive prices, so that we may secure the contract and provide constant work for all. I am also authorized by the board of directors to state that fifty per cent of the profits of the contract—mind you that amount will be comparatively small—will be apportioned as a bonus to the workpeople."

Before Captain Mostyn left Brocklington the matter was clinched, as far as the hands were concerned. There was a unanimous decision on the part of the workpeople to back up the firm, and once this step was decided upon Captain Mostyn knew that the greatest obstacle was removed, and that British grit and determination on the part of the employees would see the business through.

The Brocklington Ironworks Company sent in their tender and waited hopefully. Three months later came the news that they had secured the contract, which had been quoted at £50,500.

It was not the lowest tender. A couple of German firms were below this estimate, owing to the low value of the mark. One, the Augsburg Manufacturing Company, tendered at £30,000, while the Pfieldorf Company of Chemnitz offered to supply and deliver the material for a trifle over £20,000. The rest of the competing firms tendered considerably higher than the Brocklington Ironworks Company.

In the conditions of contract several "stiff" clauses had been inserted. The Kilba Protectorate Government required the completion of the work, including delivery, by the end of March; failing which, a heavy penalty was to be inflicted. All the steelwork was to be examined by competent inspectors, both in England and on the site, and any defective material was to be replaced at the contractors' expense plus a fine equal to the value of the replaced work.

Gallantly the employees of the Brocklington Ironworks Company rose to the occasion. Work continued at high pressure in spite of sundry difficulties. When the supply of coal ran short, the smelting furnaces were fed with wood. When the railway companies dallied over the delivery of iron-ore, a fleet of motor lorries brought the stuff direct from the pits. Once, a series of unexplained explosions destroyed a part of the rolling mills, but within a week the machinery was in working order again, and by the end of October the whole of the steelwork was ready for the Government inspectors.

It was to receive the report of the latter that Captain Mostyn had gone to town. At 8.50 Peter met his father at Trentham Regis station.

"Why, Pater," exclaimed Peter, as his sire alighted, "what have you been doing—whitewashing?"

Captain Mostyn glanced at his shoulder. His coat was speckled with white dust.

"Oh, that," he replied carelessly. "I suppose it was when some fellow working above the board-room fell through the ceiling. He wasn't hurt, only a bit scared. I'll tell you all about it later. What's more to the point, Peter, the steelwork's passed the test with flying colours, and we're shipping it on Saturday on the S.S. *West Barbican*."

"My new ship," announced Peter.

CHAPTER II
The Eavesdropper

Ludwig Schoeffer, London agent for the Pfieldorf Company of Chemnitz, was feeling at the very top of his form. He was carrying out his instructions in a manner that bid fair to be highly satisfactory both to his employers and himself, and unless untoward events disturbed the even tenor of his investigations he stood to win the sum of two hundred pounds before the day was out.

The Pfieldorf Company were both surprised and angry when the news came that their tender for work for the Kilba Protectorate had been "turned down". Their Teutonic mentality could not account for the fact that a tender considerably higher than theirs had been accepted. The war was over: why, then, should a good, old German firm be slighted and practically debarred from securing a contract that would advance the commercial prestige of the Fatherland?

At an extraordinary meeting of the directors Herr Bohme, chairman of the company, proposed a somewhat startling scheme. He suggested that the steelwork should be put in hand immediately, according to the specification of the Kilba Protectorate Government. In any case, the bridge, being of a useful design, would find a ready purchaser in one of the South American republics, or perhaps in certain parts of Africa where there was no British prejudice against German goods. The mere fact that they were picking another man's brains by copying the Protectorate's civil engineer's designs hardly entered into Herr Bohme's calculations.

"And now I come to an important proposition," continued the chairman. "It is for us to do our best to prevent this British Brocklington Ironworks Company from carrying out their contract. Somehow—how, I do not know yet—somehow that firm must be compelled to fail in their undertaking. At the critical juncture the Kilba Protectorate will be without their most important bridge, and we can well imagine the effect that will have upon the country. That is where we step in. We can offer a similar structure, complete, and in every way conformable to specification, for the sum of twenty million marks, which is ten thousand pounds more than our original tender, free on board at Hamburg. To save the situation the Protectorate Government will jump at our offer."

"But how can we prevent— —?" began one of the directors.

Von Bohme winked ponderously.

"There are ways and means, von Kessler," he interrupted. "These English fondly imagine that, now the war is over, there is no need for our admirable secret service. As you know, that organization still exists most healthily; only, instead of being the Imperial, it is now the German Commercial Secret Service."

Herr von Bohme had occasion to be vindictive towards everyone and everything British. A violent Junker, he had supported the ex-Kaiser's war policy with all his might and main, never doubting, until it was too late, of the rapid and triumphant success of the German arms. At the Armistice he had been compelled to surrender eight of his largest merchant vessels to the Allies. That practically smashed up the shipping business of which Herr Bohme was managing director. There remained the Pfieldorf Company, the activities of which bid fair to more than recoup the directors and shareholders for the loss of their mercantile marine. But von Bohme never forgot. Behind his keen business capabilities lurked the spirit of vindictiveness towards the Power that had taken so large a part in the smashing of the German Empire.

Without loss of time von Bohme telegraphed to Ludwig Schoeffer, and twenty-four hours later Ludwig presented himself at von Bohme's house in the Platz Alice at Chemnitz.

Schoeffer, although only twenty-seven years of age, had had an adventurous career. He was one of the very few German spies in England who had eluded the skilfully drawn toils of Sir Basil Thomson. At the outbreak of hostilities the spy was actually holding a British Admiralty position in Whitehall, and during the next two years he was busily serving two rival Governments at Portsmouth, Chatham, and Rosyth. At the latter place a very simple slip nearly "gave him away", and he quickly transferred his activities to the United States. There he specialized in "deferred action bombs"; ingenious contrivances detonated electrically by means of clockwork. Many a good ship owed her destruction to Ludwig Schoeffer's cunning; and, even after the cessation of hostilities, he remained in New York with the object of wrecking the ex-German vessels surrendered under the terms of the Armistice. But at last the spy was caught through the 'cuteness of a Hoboken policeman. Arrested, he was tried, found guilty, and sent for a life-sentence to Sing Sing. Three weeks later he created a record by breaking out of that grim penitentiary, and succeeded in making his way back to Germany, via San Francisco, Sydney, and Colombo.

There was nothing about Ludwig Schoeffer's appearance to betray his nationality. He might, and frequently did, pass for an Englishman, while his command of English defied detection. He was of medium height and build, dark-haired and sallow-featured. There was nothing of Teutonic stolidity about his movements. On the other hand, he walked with the elasticity and easy carriage of an Anglo-Saxon athlete.

Von Bohme received his visitor in his study, drew a thick curtain over the door, and came to the point at once.

"I want you to proceed to England, Schoeffer," he began. "Usual terms— payment by results with immediate advance to meet current expenses. You know Brocklington?"

"I was there in May and June, 1915, Herr Bohme."

"Good; but I fancy you don't know the Brocklington Ironworks."

The spy considered a few moments. To admit that he did not might be a confession of lack of local knowledge. To say that he did when he was not sure of the fact was to lay himself open to being discredited. Then he considered that perhaps his employer was trying to catch him out.

"I saw no ironworks there," he said at length.

Von Bohme grunted in satisfaction.

"For the very good reason that they came only into existence since the war. Now, read this and commit the salient facts to memory."

Von Bohme handed his caller a dossier containing the papers dealing with the Kilba Protectorate contract. There were eleven pages of closely lined typescript with marginal notes in von Bohme's own handwriting.

"You've grasped the important points? Good! Now, this is your task. Go to Brocklington, find out and report to me on the progress of the work. When necessary, shadow the directors of the Company in London. Their offices are in Chilbolton Row, off St. Mary Axe. Use every means at your disposal to hinder the work, since at all costs that steelwork must not arrive at Kilba. You understand?"

Thirty-six hours later Ludwig Schoeffer arrived at Brocklington. In the guise of a mechanic he presented himself at the works foreman's office, having previously taken the precaution of registering under the name of James Sylvester at the local Labour Exchange.

Already the contract was well in hand. Additional workmen were being taken on, and the mere fact that Jim Sylvester was a skilled riveter recommended by the local Exchange enabled the secret service agent to obtain employment forthwith.

That was all very well as far as it went, but the fact that he was actually at the works afforded Ludwig very few opportunities of getting in touch with the brains of the concern. So, after two futile attempts to hinder the work, Jim Sylvester obtained his discharge and disappeared from the neighbourhood.

By this time the spy had got to know the managing director and most of the principals by sight. His next step was to try to probe the secrets of the head office in Chilbolton Row.

Judicious inquiries resulted in the information that the Brocklington Ironworks Company's city premises were the ground floor of a large, somewhat dingy building. The second and third floors were occupied by shipping agents; the first floor was at present unoccupied.

Three days later Ludwig Schoeffer was in possession of the hitherto vacant rooms immediately over the Brocklington Ironworks Company's offices, but not as Ludwig Schoeffer. A card affixed to the door announced to anyone who had occasion to visit the upstairs offices that Mr. Josiah Sherringham, London agent for Messrs. Hoogenveen, bulb growers, of Haarlem, would be in attendance daily from ten to four. Since Messrs Hoogenveen, had no material existence, it was extremely unlikely that clients would call upon Mr. Josiah Sherringham. Nor did the tenant of the first floor want any. Usually the door was locked, generally from the outside, and inside whenever the directors of the Brocklington Ironworks Company held converse in the room below.

Amongst Mr. Josiah Sherringham's office furniture was a dictaphone, the mouthpiece of which was extended by means of a length of india-rubber tube and rested above a hole in the ceiling of the room below. Some years previously the premises had been renovated and electric light installed in place of gas, but the huge ornamental rose from which a chandelier once depended formed a convenient camouflage for the eavesdropper's operations.

Whenever the directors of Brocklington Ironworks Company held a board meeting, Ludwig Schoeffer was an unseen listener. Being rather particular about his appearance the spy invariably donned a suit of workmen's overalls, lest his clothes should show signs of having come in contact with the dusty floor. Fortnightly, transcribed records of the British firm's progress were transmitted to the Platz Alice at Chemnitz.

At length came the momentous meeting at which Captain Mostyn was to announce the result of the Kilba Protectorate Government's inspector's preliminary tests of the steelwork; and also the arrangements made for the shipment of the material to its destination.

The dictaphone was purring softly. Ludwig, on his hands and knees, had prized up some floor-boards and was listening to the report. In his eagerness he could not wait for the wax cylinders to tell him what was transpiring.

At a critical moment the dictaphone ceased functioning. The eavesdropper half rose to attend to the instrument. His knees slipped on the narrow joists, and the next instant, amidst a rending of laths and plaster, he landed on his back upon the table around which were seated the directors of the Brocklington Ironworks Company.

CHAPTER III
Reporting for Duty

"Now, Pater, tell me how you got on in town," prompted Peter.

"Famously! The inspector's report laid special emphasis upon the excellence of the castings, and I've no doubt that the final tests will be equally successful. We also secured very reasonable freightage. The *West Barbican* is not a fast vessel—fifteen knots is, I believe, her limit—but she will be able to deliver the goods well in advance of the time specified. It is certainly remarkable, Peter, that you should have been appointed to that same craft."

"I'm jolly glad," replied Peter. "It's about time I went afloat again. It looks as if I'm giving this winter a miss, eh? By the by, didn't you say something about a fellow tumbling through the ceiling?"

Captain Mostyn laughed.

"Yes, it was very funny," he replied. "We were all deep in business when there was a jolly old crash, and before we realized it there was a man—a workman—spread-eagled on the table. Winterton and Forsyth helped him up and asked if he were hurt. ''Urt?' he remarked bitterly. 'Not 'arf.' But he was able to walk without assistance. It seems that he was engaged in overhauling the electric-light fittings in the office over ours, and something carried away and let him down. It might have been worse.... Have you your kit ready?"

"Almost," was the rejoinder. "I'll have to go up to town on Wednesday, because my tropical outfit wants renewing. So we're to run round to Brocklington?"

"Yes," replied Captain Mostyn. "We've made arrangements for the steelwork to be shipped from there. Saves a lot of trouble sending it to East India Docks. We gain on the estimate that way, although, of course, we are practically chartering the *West Barbican* for two or three days."

At ten on the following Thursday Peter Mostyn boarded the *West Barbican*. The ship was of about 7000 tons, single funnelled, and with two stumpy masts with telescopic topmasts and a sheaf of derricks to each. She was still coaling and her decks were deep in grimy dust. With the exception of the officers the ship was manned by lascars—a novelty so far as Peter was concerned.

A burly, jovial-featured man in a grimy uniform, and wearing a muffler under the turned-up collar of his tunic, greeted Mostyn as he stepped off the gangplank.

"Hello, you're our Sparks, aren't you?" inquired the man. "My name's Preston when it's not Salthorse. Just now it ought to be Coaldust. I'll take you along to see the Old Man, and, when he's done with you, come to me for the keys of the wireless cabin. I'm Acting Chief."

Picking his way between coal-bags, dodging knots of bizarrely clad lascars, who with shrill cries dragged the sacks of fuel to the bunker shoots, Peter followed the Acting Chief Officer to the for'ard end of the boat-deck, where the skipper of the S.S. *West Barbican* had his cabin. Over the jalousied door was a brass plate with the word "Captain"; just below the plate was a card on which appeared, in bold and rather straggling handwriting, the intimation: "Don't knock—come in."

"Carry on, old son," urged Preston—and left Peter to his own devices.

For a brief instant Peter hesitated. Then, force of habit gaining the ascendancy, he knocked discreetly upon the white-enamelled door.

"What are you hanging on to the slack for?" demanded a bull voice. "Where are your blessed deadlights? Can't you read?"

The Wireless Officer opened the door and stepped briskly into the cabin.

Sitting in an arm-chair in front of a table littered with books and papers was a short, thick-set, bearded man. He was in his shirt-sleeves; a salt-stained uniform cap was perched on the back of his head, leaving exposed a wide, vein-traced forehead bordered on either side by closely cropped grey hair. His complexion was a dusky red, while his choleric blue eyes peered beneath a pair of beetling bushy eyebrows.

This was Mostyn's first impression of Captain Antonius Bullock, master of the good ship *West Barbican*.

"No doubt his bark is worse than his bite," soliloquized Peter, then, aloud, he said:

"I wish to report for duty, sir."

"Another time you come into my cabin do as you're told," growled the Old Man. "Can't waste my breath telling people to come in—may want it badly some day. Where's your permanent discharge book?"

Mostyn had the article ready to hand—one of those thin, blue-covered booklets which, according to Board of Trade Regulations, must be in the possession of every officer and man of the British Mercantile Marine. It is his passport through life as long as he remains under the Red Ensign, and corresponds with the parchment certificate of the Royal Navy.

"'Report of character: for ability, very good; for general conduct, very good'," read the Old Man aloud. "Let's hope that'll continue. Hello! what's this: last ship the *Donibristle*. I hope I haven't shipped a Jonah."

"I hope not too, sir," agreed Mostyn.

"Carry on, then," was the brief rejoinder, and the introductory interview terminated.

Truth to tell, Captain Antonius Bullock was not particularly fond of wireless operators. This antipathy was not due to the individual but to the system. Although wireless officers came under the captain's orders for disciplinary purposes, they were governed by the rules and regulations of the wireless company who employed them. Consequently it was possible, and often probable, that the Old Man might issue an order to the radio staff that ran directly counter to the wireless regulations; and, if the skipper were short-tempered and disinclined to listen to explanations, matters would come to a climax by the wireless officer flatly but firmly declining to carry out the Old Man's behests.

On the previous voyage such an incident had actually occurred. Captain Bullock had given an impossible order—impossible according to the wireless operator's reading of the regulations. The Old Man lost his temper and told the operator to work double watches for the rest of the voyage; the latter retaliated by "logging" the skipper. This drastic step rather frightened the choleric Bullock, especially when, on further consideration, he found that he was in the wrong. Before the *West Barbican* arrived in London River, skipper and wireless operator had a private and amicable conversation, with the result that the latter expunged the offending record from the log. But the matter still rankled in Captain Antonius Bullock's broad bosom, and, since he could not consign the system to perdition, he vented his resentment upon the wireless officers under his command.

There was no denying Captain Bullock's qualifications as a seaman. He was courageous, resourceful, skilful, and, withal, cautious. He had been at sea for more than thirty-five years, having served his apprenticeship in a square-rigged ship and worked his way up through that roughest of rough schools—the South American cattle-boats—to his present responsible position of senior captain of the Blue Crescent Line.

Outside the captain's cabin Peter was met by a tall, slim Hindustani wearing a blue dungaree suit, a pair of straw-plaited shoes, and a red "pill-box" hat.

With Oriental obeisance, yet not without a certain display of dignity, the "boy" salaamed.

"Me Mahmed, sahib. Me you boy," he announced.

Peter regarded his new acquaintance critically. Mahmed was a Madrasi of about twenty years of age, with features handsome in an Oriental way. In spite of his weird attire—for during coaling operations the native crew had discarded their smart but serviceable uniforms—there was something about the youth that impressed his new master favourably.

"Want *char*, sahib?"

The word "char" was not a stranger to Peter Mostyn. Of Eastern derivation, and meaning "tea", it has been adopted by Britons in all quarters of the globe; and even in Flanders and the north of France peasants have learned the word.

Receiving an affirmative reply, Mahmed glided noiselessly away, while Peter set out to find the Acting Chief Officer and obtain the keys of the wireless room.

"So the Old Man hasn't chawed you up?" remarked Preston, with a broad grin. "He's not a bad old lad when you know him. What's your name?"

Peter enlightened him.

"Dash it all!" exclaimed the Acting Chief. "I've heard of you, young fellah-me-lad! Weren't you in that *Donibristle* stunt? We've shipped a *pukka* hero this trip."

"Don't know about that," protested Peter. "The Old Man has just told me I'm a Jonah."

CHAPTER IV
The Greenhorns

Armed with a bunch of keys, Peter made his way up several ladders until he gained the box-like structure bearing a brass plate inscribed "Wireless Cabin".

The erection was of solid construction, lighted by six brass-rimmed scuttles. The door, opening aft, was affording support to a couple of pale-faced, weedy-looking youths, who, on seeing Mostyn appear, made no attempt to shift their position, not even to the extent of removing their hands from their pockets.

The Wireless Officer realized at once who these lads were. Already he had had his suspicions on the point. The fact that he had received no intimation of the presence of a junior wireless operator rather prepared him for the discovery.

"What are you doing here?" he demanded.

The taller of the two boys glanced at his companion as if urging him to reply. Receiving no encouragement from that direction he gazed vacantly into space.

"Bloke dahn there told us to 'ang on 'ere," he announced, in the sing-song voice of a city-bred, elementary schoolboy.

"We're Watchers," added his companion.

"Oh, are you?" rejoined Peter. "Then please to remember that when you are spoken to by an officer you will address him as 'sir'."

Mostyn was not snobbish—far from it, but the attitude and tone of the pair went against the grain. It was the first time that he had found himself "up against" the genus Watcher, and the impression served to support the adverse reports he had heard of the general incompetence and uselessness of the class.

"Watchers" were the outcome of an ill-advised step on the part of shipowners towards economy. A second-class ship, such as the *West Barbican*, might carry either two trained and Government-certificated operators—men who were qualified in both the practical and technical side of radiography—or she might carry one operator and two Watchers.

The latter were simply and solely unskilled youths who were sent on board ship to "listen-in" for wireless messages. They took turns in putting on the telephones and waiting for wireless calls. All they could do—or were expected to do—was to recognize two call signals: the SOS and TTT, the latter an urgent general signal of lesser importance than the well-known call for aid. To the Watchers the Morse Code was a sealed book. Their occupation was of a blind-alley nature. They could hardly hope to qualify as operators, lacking the aptitude, intelligence, and opportunities for gaining their wireless ticket. In short, they were a cheap product whereby their employers sought to cut down expenses by dispensing with one of two wireless officers, regardless of the grave risk that an error on the part of these half-baked dabblers in radiography might endanger the ship.

As a class, too, they were resented by the wireless staff proper. Not only would the employment of Watchers tend to diminish the numbers of *pukka* wireless officers serving afloat; but the wireless officer on a ship carrying Watchers would be always on duty although not actually in the cabin. Instead of taking "tricks" with his "opposite number" he would be liable to be summoned by the Watchers on duty at any hour of the day or night, simply because his assistant could not, and would not be allowed to, receive or send out messages.

"Is this your first voyage?" asked Peter, addressing the taller Watcher.

"Yes," was the reply.

"Yes, what?" demanded Mostyn sharply.

"Yes, sir."

"That's better," continued Peter, as he unlocked the door, the two lads having summoned up enough physical energy to stand aside. "What's your name?"

"Partridge,"—pause—"sir."

"And yours?"

"Plover, sir."

"Weird birds," soliloquized Mostyn; "but perhaps they'll lick into shape."

His first impression of the interior of the cabin was not a good one. The *West Barbican* had been laid up for nearly four months, and, although her late Sparks had conscientiously carried out his written instructions as to the precautions to be taken when "packing up", the prolonged period of idleness had not improved the appearance of the apparatus. In spite of a liberal coating of vaseline the brasswork was mottled with verdigris; moisture covered the ebonite and vulcanite keys; the roof had been leaking, the course of the water being indicated by a trail of iron rust upon the white paint.

Dust covered everything, while the absence of fresh air, owing to the scuttles having been secured for months, was distressingly noticeable.

"Phew! What a reek!" exclaimed Peter, stepping backwards into the open and nearly colliding with the impassive Mahmed.

"Char, sahib."

Mostyn gulped down the hot beverage, and literally girded up his loins for direct action.

"Nip below," he ordered, addressing the still torpid Partridge. "Get hold of a bucket of hot water, a squeegee, and some swabs. Look lively, Plover; get busy with those scuttles. Open all of them. Scuttles, man; those round glass windows, if you like."

Watcher Plover tackled his allotted task with a zest that rather surprised his superior officer, but it was not until five minutes later that Peter found the Watcher trying to unbolt the brass rims instead of unthreading the locking screw.

"Belay there," exclaimed Mostyn. "Don't take the whole of the cabin down. Let me show — —"

His words were interrupted by a metallic clatter followed by sounds of falling water. Watcher Partridge's hob-nailed boots had slipped on the brass treads of the ladder, and he had finished up ingloriously upon the deck, sprawling upon his back in a puddle of coal-grimed water.

While the unlucky Partridge was making a prolonged change and refit, Mostyn with his other assistant tackled the demon dirt in his lair. Not until the dust was removed and the paint-work and floor well scrubbed and dried did Peter begin to overhaul the "set".

The dull daylight faded and gave place to night, but still the indefatigable wireless operator carried on, until the bell summoning the officers to dinner warned him that it was time to knock off.

"Not so bad," he conceded modestly, as he surveyed the array of glittering brasswork and polished vulcanite. "I'll leave the actual tuning up and testing till to-morrow. Buzz off, you fellows. You won't be wanted until two bells in the forenoon watch."

Locking the door, Mostyn made his way to his own quarters. His cabin was of the usual double-berth type, one bunk being superimposed immediately above the other. In this instance he was the sole occupant of the cabin, and rather grimly he commented upon the saying that it's an ill wind that blows nobody any good. Had he not been called upon to endure Messrs. Partridge and Plover, he would have had to the share cramped quarters with another wireless officer.

In the adjoining cabins the jaded occupants were busily engaged in removing the traces left by their arduous labours. The coaling operation had been completed. The bunkers had been trimmed, decks washed down, and the hideous but necessary coaling-screens stowed away. Yet the ship reeked of coal-dust. The alleyways seemed stiff with it. It penetrated even into the locked and carefully curtained cabins and saloons.

On board the S.S. *West Barbican* there was nothing in the way of formal introduction. A newly joined officer simply "blew in" and made himself at home. When off duty the fellows were more like a pack of jolly schoolboys than men on whose shoulders rested a tremendous weight of responsibility. They accepted a newcomer as one of themselves, and, unless he were an out-and-out bounder, soon set him entirely at his ease.

In vain Peter scanned the features of his new shipmates in the hope of recognizing a familiar face. For the most part the officers had been on board for lengthy periods, the interval of idleness notwithstanding. They were a conservative crowd in the Blue Crescent Line, and, since Mostyn had served on vessels plying between Vancouver, Japan, and China, he was not surprised, although disappointed, to find that his hopes were not realized.

"Have we got our orders yet?" inquired the Chief Engineer, addressing the Acting Chief Officer, who, in the absence of the skipper, was sitting at the head of the long table.

"Yes," replied Preston. "We're off to a place called Brocklington, on the East Coast, to pick up the bulk of our cargo—steelwork, worse luck. Next to iron ore I know of nothing worse. It'll make the old hooker roll like a barrel. After that we return to Gravesend on Monday, pick up our passengers, and then away down Channel. Let's hope we don't see London River again until shipping looks up considerably. I've had enough of kicking my heels on the beach, and I guess you have too. Once we go East the owners aren't likely to send us home in ballast."

"Dull times these, especially after the war," remarked Anstey, the Third Officer. "Even those pirate stunts in the Atlantic and Pacific are a wash-out."

"Which reminds me," added Preston, indicating the modest Mostyn. "Our Sparks here was in the *Donibristle* when that Porfirio blighter collared her. For first-hand information apply to our young friend here."

So Peter had to relate briefly the hazardous adventures of the crew of his former ship, after they had been taken into captivity by the swashbuckling pirate Ramon Porfirio. Before the evening was over he felt as if he had known his new messmates for ages.

CHAPTER V
Under Way

Mostyn awoke soon after daybreak, or rather was aroused by the appearance of Mahmed with a cup of *char* in one hand and a copper jug full of hot water in the other.

It was a novel experience for Peter to watch the deft movements of his servant, who seemed to possess an uncanny knowledge of where his master's personal belongings were stowed. Mostyn's safety-razor, strop, shaving-pot, and soap were placed ready for use; his boots were shining with unusual brilliancy, even in the comparatively feeble rays of the electric lamp. His clothes, folded and pressed, were placed ready to put on. How and when Mahmed had contrived to make these preparations without disturbing his master rather puzzled the Wireless Officer, for he considered himself a light sleeper.

Breakfast was more or less a scrambled affair, many of the officers having to gulp down a cup of hot tea and hurry off to their appointed tasks, for the *West Barbican* was sailing at noon, and there were multitudinous duties to be seen to before the ship was actually under way.

Directly after breakfast Peter hastened to the wireless cabin in order to put in an hour's uninterrupted work before the appearance of his two inefficient assistants. Not that they would have worried him by asking questions, intelligent or otherwise. It was their wooden-faced passivity that Peter found disturbing. He wondered by what manner of means such a quaint pair of birds was taken into the Company's service.

At four bells—ten o'clock—Mostyn had got his set into working order, and a quarter of an hour later the wireless inspector came on board to receive the radio-officer's report, and to satisfy himself that the installation was in every way efficient.

"I can give your little outfit a clean bill of health pretty quickly, Mr. Mostyn," remarked the inspector. "Evidently your predecessor left you very little to do. Once you've broken in your two Watchers you ought to have a very soft time."

"I hope so," rejoined Peter guardedly, but he had grave doubts on the subject. Not that he wanted a "very soft time"—he was far too energetic for that—but because he felt convinced that his assistants were not cut out for the job.

At length a blast on the siren announced that the West Barbican was about to leave the dock. Peter left the cabin to watch the now familiar yet engrossing scene, familiar save for the fact that for the first time he had shipped with a crew of lascars. It was a strange sight to see the natives on the fo'c'sle, carrying out orders under the serang, and to watch a barefooted lascar go aloft, gripping the shrouds with hands and toes with equal facility.

Under the gentle yet firm persuasion of a couple of fussy tugs the West Barbican renewed her acquaintance with London River. There were no demonstrations at her departure. None of the officers had any relations or friends to wish them God-speed from the shore, and, since the passengers had not yet embarked, the usual display of farewells was not in evidence.

It was not until the ship entered Sea Reach that Peter called his assistants.

"You, Partridge, will take on now," he said. "Plover, it's your watch below. You'd better see that you get some sleep. Now, you know your duties, Partridge?"

"Yes, sir."

"Right-o; carry on!"

Partridge sat down and clipped on the telephones. Peter left him, but promised himself to visit the cabin pretty frequently, to see that the Watcher was watching. Meanwhile he had plenty to do in the clerical line, filling up forms and making reports upon various technical matters.

Half an hour later Mostyn returned to the wireless-room. He was not surprised to find that Master Partridge was lying on the floor, having previously "mustered his bag" with the utmost impartiality. Watcher No. 1 was down and out.

"The poor bounder can't help being sea-sick, but he ought to have been a little more considerate," soliloquized Mostyn, after he had told the unhappy Watcher to clear out and turn in. In fact, Partridge was so bad that Peter had to assist him down the ladder until he handed him over to the care of a lascar.

Although the ship had not yet passed the Nore she was rolling considerably, for there was a fresh wind on the starboard beam. Evidently she was doing her best to live up to her reputation. But Peter made light of the motion. With the telephones clipped to his head he sat in the open doorway of his "dog-box", watching the ever-changing seascape so far as a couple of boats in davits permitted.

When the hour arrived for Watcher Plover to take over the watch, that individual was not forthcoming. Peter waited a full ten minutes and then told a *seedee-boy* to warn the absentee.

Presently the Indian messenger returned with a faint trace of a smile on his olivine features.

"No go, sahib," he announced. "He ill—very sick like to die."

Mostyn shrugged his shoulders and "carried on". Fortunately he had had a fairly good night's rest. The treble trick he could endure with equanimity, buoyed up by the hope that the indisposition of his two inefficient assistants would be of short duration, especially as the *West Barbican* was due to berth in Brocklington Dock by six the next morning.

Before long the weather began to get decidedly dirty. The haze that had been hanging over the coast had vanished, but to the east'ard banks of ragged-edged indigo-coloured clouds betokened a hard blow before very long. The wind, too, had backed from sou'-sou'-east to nor'-nor'-east, and was rapidly increasing in force.

The *West Barbican* was not belying her reputation for rolling. In the wireless cabin, between forty and fifty feet above the sea, everything of a movable nature was slithering to and fro with each long-drawn oscillation of the ship. More than once Peter had to grip the table to prevent his chair sliding bodily across the deck. The wind was thrumming through the shrouds, and whistling through the still open scuttles, while the aerial vibrated like a tuning fork in the shrieking blast.

It was one of those sudden gales that play havoc with small craft, especially in the comparatively shallow waters of the North Sea; but, although Peter kept a vigilant look out for SOS signals, the air was remarkably free from radio calls. At intervals he could hear a peculiar buzzing in the ear-pieces—a noise that he knew from previous experience to be distant rain.

A shadow darkened the cabin. Peter turned his head and saw Anstey, the Third Officer, standing in the doorway. He was prepared for the storm, his head being partly concealed by a sou'wester, while a long oilskin coat and a pair of india-rubber boots completed the visible portion of his rig-out.

"Hello, Sparks!" he exclaimed. "How goes it? Anything doing?"

"Absolutely nothing," replied Mostyn. "Everything's as quiet as the proverbial lamb. I suppose——"

He broke off suddenly.

Anstey made some remark, but the Wireless Officer took not the slightest notice. Already he had snatched up a pencil and was scribbling upon the ever-ready pad.

It was a TTT or urgent warning signal. Mostyn wrote it down mechanically without knowing its import, but the Third Officer, looking over Peter's shoulder, made a grimace as he deciphered the other's scrawl:

"CQ de GNF—TTT—mine warning—S.S. two-step reports 1630 sighting two mines, lat. 53° 20' 15", long. 1° 5' 30" east stop mines just awash barnacle covered apparently connected by hawser—end of message."

"By Jove!" exclaimed Anstey. "Just our luck. Right in our course, an' it's my blessed watch."

CHAPTER VI
A Night of Peril

Making his way to the chartroom the Third Officer "laid off" the position of the mines. His rough guess proved to be remarkably accurate. According to the position given, the source of danger was only a few miles from the Outer Dowsing Lightship, and the *West Barbican* had to pass close to the Outer Dowsing on her course to Brocklington.

Anstey's next step was to inform the Captain. The Old Man, a sailor to the backbone, was in the chart-house in a trice, where, after a brief but careful survey of tide-tables and current-drift charts, he was able to determine the approximate position of the floating mines when the ship would be in the immediate vicinity of the light-vessel. Allowing for the set and strength of the tide and the drift caused by the wind, between the time the mines were first sighted and the time when the *West Barbican* entered the danger-zone, he was able to assert that, if the ship's original course were maintained, she would pass at least ten miles to the east'ard of those most undesirable derelicts.

"I think we're O.K., Mr. Anstey," he remarked. "Besides, for all we know the mines might have been exploded by this time. Those naval Johnnies are pretty smart at that sort of thing. Well, carry on. Let me know if there are any supplementary warnings."

The Old Man returned to his cabin, and was soon deep in the pages of a novel; while Anstey resumed his trick, thanking his lucky stars that, unlike Mostyn's, his watch was not indefinitely prolonged through the shortcomings of two sea-sick "birds".

Just as darkness set in, the gale was at its height. Clouds of spray flew over the bridge as the old hooker wallowed and nosed her way through the steep, crested waves, for the wind had backed still more and was now dead in her teeth.

Even in the wireless-cabin the noise was terrific. The boats in davits were creaking and groaning, as they strained against their gripes with each disconcerting jerk of the ship. Spray in sheets rattled upon the tightly stretched boat-covers like volleys of small shot, while the monotonous clank-clank of the steam steering-gear, as the *secuni* (native quartermaster) strove to keep the ship within half a degree of her course, added to the turmoil that penetrated the four steel walls of the cabin.

Vainly Peter tried to concentrate his thoughts on a book. Yet, in spite of the fact that he was wearing telephones clipped to his ears, the hideous clamour refused to be suppressed. Reading under these conditions was out of the question. He put away the book and remained keeping his weary watch, valiantly combating an almost overwhelming desire for sleep.

Suddenly, with a terrific crash, something hit the deck of the flying-bridge immediately above the wireless-cabin. For a moment Peter was under the impression that one of the foremost derricks had carried away and crashed athwart the roof of the cabin.

Soon he discovered the actual cause. The stout wire halliard taking the for'ard end of the aerial had parted, and the two wires, spreaders, and insulators had fallen on the boat-deck.

Removing the now useless telephones and donning his pilot coat, Mostyn went out into the open, glad of the slight protection from the cutting wind afforded by the canvas bridge-screens and dodgers. Already lascars, in obedience to the shrill shouts of the serang and *tindal* (native petty officer), had swarmed upon the bridge ready to clear away the debris.

Accompanied by the bos'un Mostyn made a hasty examination of the damage. The aerials had fortunately fallen clear of the funnel, and, although the for'ard insulators had been shattered, the drag of the wires had kept the after ones from being dashed against the main topmast.

It was "up to" the Wireless Officer to repair and set up the aerials as soon as possible.

While the lascars were clearing away a spare halliard, Peter began to replace the broken spreader and its insulators. Cut by the keen wind, drenched with the rain and spray, and chilled to the bone in spite of his heavy pilot coat, Mostyn struggled with refractory wires until his benumbed hands were almost raw and hardly capable of getting a grip on the pliers.

It was a hit-or-miss operation. In the circumstances he had no means of testing the insularity of the aerial. He could only hope that, when once more aloft, it would function properly.

With a sigh of relief he completed the final splice and turned to the serang.

"Heave away!" he ordered.

The man gave a shrill order. Instantly the hitherto passive line of lascars handling the slack of the rope broke into activity. Gradually the aerial tautened, as a score of brown-faced, thin-limbed natives tailed on to the hauling part of the wire halliard. Quickly at first, then with gradually diminishing speed, the double line of wire rose from the deck and disappeared from view in the spray-laden darkness of the night, and presently the serang reported that the aerial was close up.

Mostyn returned to his post. Glancing at the clock he noted with astonishment that the task had taken him exactly an hour. Then, replacing the telephones to his ears, he endeavoured to thaw his benumbed fingers in front of the electric-light globe.

Hour after hour passed in monotonous inactivity. The appearance of the devoted Mahmed with a cup of tea and a plate of sandwiches—most of the tea was spilt, and the sandwiches were abundantly salted and moistened in the process of mounting the bridge—proved a welcome diversion.

Just before midnight a second disaster occurred to the aerial. This time the double wires parted, practically simultaneously, about midway between the masts. This point, being almost immediately above the funnel, is always a fruitful source of trouble, owing to the comparatively rapid deterioration set up by the gases from the furnaces.

Repairs, even of a makeshift nature, were for the present out of the question. It was impossible to send men aloft to assist in setting up the wires. No human being could hold on in such a gale, far less perform the intricate task of reeving fresh halliards and wires. All Mostyn could do was to make all secure in the wireless-cabin. He was then free to turn in and enjoy a few hours' rest, until the ship's arrival at Brocklington Dock should afford an opportunity for repairing the damage.

Peter was exchanging a few words with the officer of the watch when the attention of both was attracted by a flash.

"Distress signal!" exclaimed Peter.

"Not vivid enough," rejoined his companion "Might be a rocket from one of the Dowsings—the Inner, most likely. If— —"

Another flash, faintly visible through the murk, interrupted Anstey's words. For several seconds both men listened intently for the double detonation. None was audible. Distance and the howling of the elements had completely deadened the reports.

Even as they looked a steady pin-prick of reddish light appeared on exactly the same bearing as the previous flashes. For perhaps fifteen seconds it remained constant; then momentarily it grew in volume until a trailing column of ruddy flame, fringed by a wind-torn cloud of smoke, illuminated the distant horizon.

Bringing his night-glasses to bear upon the source of the flames the Third Officer studied the scene. Then, replacing the binoculars, he shouted to his companion:

"Vessel ablaze from end to end. Tanker, I guess. I'm off to call the Old Man."

Captain Bullock was quickly out of his cabin. He had waited merely to put on his bridge-coat over his pyjamas and thrust his bare feet into a huge pair of sea-boots. He was one of those powerfully framed, tough men for whom the sudden change of temperature had no terrors and few discomforts.

Shouting a hoarse yet unmistakable order to the secum at the wheel, and ringing down to the engine-room for increased speed, Captain Bullock waited until the *West Barbican* had steadied on her new course, then he turned to the Third Officer.

"She's a tanker, right enough, Anstey. Got it properly in the neck. See that the boats are cleared away, although I'm afraid there's precious little chance of using them in this sea. I'm off to shift into thicker togs."

In five minutes the Old Man returned. By this time the *West Barbican*, making a good twelve and a half knots against the head wind and sea, had got within a couple of miles of the doomed vessel.

Already she was well down by the head, and blazing furiously from stem to stern. To windward of her the seas were breaking heavily against the hull of the burning ship. Already she had lost way and was drifting broadside on to the wind. Cascades of water pouring over her listing deck had no effect in quenching the flames but merely raised enormous clouds of steam to mingle with the flame-tinged, oily smoke. To leeward the sea was calm for almost a mile, owing to the liberation of the oil. And not only was it calm: it was a placid lake of fire, as the floating, highly inflammable coating of petroleum burnt furiously in half a dozen detached areas.

"See any signs of a boat?" demanded the Old Man.

"No, sir," replied Anstey.

"Thought not," was the rejoinder. "A boat would be swamped to wind'ard, and burnt to a cinder to lee'ard. Doubt even whether the poor fellows had a chance to lower away — — What's that on our port bow? By heavens, Anstey, it's a boat!"

Both men levelled their binoculars. Mostyn, keeping discreetly in the background, made use of the chartroom telescope.

Silhouetted against the glare was a ship's boat. There were people in her, but they were making no apparent effort to draw away from the danger zone. Rising and falling on the long, oily swell, the frail craft was midway between two patches of fiercely burning oil that threatened to converge and destroy the boat and its human freight.

"We'll have to risk it, Anstey," decided the Old Man, as he rang for half speed. "I only hope the lascars'll stick it. I'm going to take the old hooker between those patches of burning oil. We'll try towing the boat clear. If that fails we'll have to lower one of our own boats. Pass the word for the serang to stand by to heave a line, and then give an eye to the *secuni*. If he runs the ship into either of those patches it'll be a serious matter."

"Ay, ay, sir."

Ringing for stop, Captain Bullock knew that there was sufficient way upon the ship to enable her to close the boat without the former being out of control. Allowance had also to be made for the wind, which, owing to the alteration of course, was now two points on the starboard bow.

The heat was now quite perceptible, while at intervals wisps of black, suffocating smoke swept to lee'ard, completely enveloping the *West Barbican*. On either side of her were expanses of burning oil, bubbling and popping in a series of miniature explosions, as the heated water beneath the oil vapourized and blew out through the covering layer of burning viscous liquid.

Right in the centre of the steadily decreasing avenue of unlighted oil lay the boat. Two cables' lengths beyond, and now a glowing mass of white-hot metal, lay the burning tanker, awash for'ard and with her propeller showing clear above the agitated water.

Admirably manoeuvred and conned by the Old Man, the *West Barbican* drew near the tanker's boat. Slowly she passed within heaving distance. The now excited lascars heaved lines, several of which fell short. Two at least dropped athwart the boat, but no attempt was made on the part of her crew to secure them. The luckless men were either dead or else rendered insensible by the hot, suffocating air.

The ship had now lost way. Her head was beginning to pay off. It was necessary to go ahead in order to regain steerage way; but, at the same time, if the work of rescue were to be consummated, it would be necessary to make use of one of the *West Barbican's* boats.

"Lower away!" roared the Old Man.

At that moment the tanker disappeared beneath the surface. The tower of flame that enveloped her died down to a mere flicker, completely outclassed by the glare of a dozen distinct patches of fiercely burning oil.

The lascars manning the falls hesitated, while their comrades in the boat showed signs of panic. In the confusion they noticed that, unaccountably there was no officer on board the lifeboat.

Mostyn was one of those men who in moments of danger are prone to act independently—they simply cannot remain passive spectators when there is work to be done. It was no business of the Wireless Officer to go away in the boats. His duty was to stay by the wireless gear. But in this case Peter knew that he could do nothing in the cabin with the aerial out of action. He could be of use in the boat, to take command and steady the decidedly "jumpy" Asiatics.

The overwhelming instinct to bear a hand seized him in an instant. Running aft to where the lifeboat swung outboard he leapt into the stern-sheets, grasped the yoke lines, and shouted to the tindal to lower away. The man, seeing that a sahib was in the boat but not recognizing who he was, gave the word to the lascars manning the falls, and the boat was lowered rapidly and evenly.

Mostyn had a momentary vision of the lighted scuttles slipping upwards as the boat dropped down past the ship's side. Then with a sharp flop the lifeboat struck the oily surface. Simultaneously the lower blocks of the falls disengaged, and the boat began to drift astern.

"Give way!" ordered Peter.

The lascars, trained to obey commands issued in English, acted smartly. With the presence of a sahib in the lifeboat their fears, if not entirely banished, were cloaked by the sense of discipline.

"Pull starboard; back port."

The lifeboat turned in almost her own length.

Already the steadily converging patches of flames justified this order. To turn under the use of the helm alone would bring the boat in contact with the oil-fired water.

"Together—way 'nough—in bow."

In five minutes from the time Peter had taken his place in the stern-sheets the two boats were gunwale to gunwale. In the tanker's whaler were seven human forms huddled in weird postures, either on the bottom-boards or across the thwarts.

Whether they were dead or alive Mostyn knew not. All he could do was to have the seemingly inanimate bodies transhipped, and then return to the *West Barbican*—if he could.

Working like men possessed, four of the lascars unceremoniously bundled the bodies into the lifeboat. Then, pushing off, they resumed their oars, pulling desperately for the ship, which was now gathering sternway at a distance of a cable's length.

For the first time Mostyn realized the extreme gravity of the situation. The ship was now gathering sternway, drifting rapidly to lee'ard the while. The churning of her propeller had caused a large patch of burning oil to still further contract the narrow fairway between the ship and the boat.

Peter knew full well that he and the boat's crew stood less than a dog's chance should the fiery sea cut them off. He was also aware of the great difficulty of being picked up by the ship, since the latter had herself to be constantly manoeuvring to avoid contact with the fire. Even if the lifeboat escaped the flames, there arose the danger of her being crushed by her parent. In that case there would be little or no chance of swimming in the thick layer of oil that had not as yet become ignited.

It was touch and go. Dazzled by the glare, partly stifled by the thick smoke, and scorched by the hot, raging wind, Peter all but lost his bearings. A momentary dispersal of the smoke showed him the hull of the *West Barbican* less than four boats' lengths away.

"Boat oars!"

The now thoroughly scared lascars obeyed very hurriedly. The bowman grasped and engaged the for'ard falls, pulping one of his fingers in the operation. Almost simultaneously the lower block of the after falls was hooked on, and with a disconcerting jerk the lifeboat rose clear of the water.

Only by a few seconds had she won through. Before the boat was hoisted home the sea beneath her was covered with crackling, spluttering flames.

CHAPTER VII
"Logged"

Peter Mostyn's chief desire upon regaining the deck was to go below and get something to drink. Now that the immediate danger was over, his throat was burning like a lime-kiln, and his head was buzzing as if he had taken an overdose of quinine.

Slipping off his lifebelt—he had donned it mechanically on rushing to the boat, although in the circumstances the advantages of wearing a lifebelt were of a negative order—Peter returned to the bridge, keeping discreetly in the background.

The Old Man was fighting a tough battle. With Preston and Anstey he was extricating his command from a perilous situation, where skilful seamanship alone could regain control of the helm without allowing the vessel to wallow helplessly in the fiery sea. Putting the ship ahead and astern alternately the Old Man allowed her head to pay off under the force of the wind until he saw a chance of turning. Then, with a grunt of supreme satisfaction, he rang for full speed ahead. Five minutes later the *West Barbican*, clear of the oil-calmed water, was rolling in the tempestuous seas.

"Carry on, Mr. Anstey," he ordered. "Lay her on her old course."

He turned abruptly on his heel, intending to see how the survivors of the tanker were faring. As he swung round he noticed Peter standing under the lee of the wireless cabin.

"Mr. Mostyn!"

"Sir?"

"How many survivors?"

Peter told him.

"A smart bit of work of yours, Mr. Mostyn, but—oh, very well, go below and turn in. I'll see you in the morning."

The Wireless Officer obeyed only too gladly. As he washed the grime from his face he reflected that, thanks to the damaged aerial, he would have an uninterrupted watch below.

For a long time he lay awake in his bunk. It was not the heavy rolling that was responsible for his sleeplessness. The whole of the night's adventure passed in review, its horrors intensified in retrospect. It was not until dawn was breaking that he fell into a fitful slumber.

Meanwhile the skipper had his hands full. In the absence of a doctor he and the purser were attending to the helpless survivors of the tanker. Of the seven removed from the boat only two were conscious, and one of the pair had a compound fracture of the right leg.

His companion was able to give an account of the disaster. The vessel was the American-owned oil-tanker *Bivalve* of and from New York for Hull. She had struck the two drifting mines, concerning the presence of which a general wireless message had been sent out. Both exploded amidships, one on either side, about fifty feet for'ard of the engine-room, which in vessels of the *Bivalve's* type are well aft. Within a few minutes the petroleum tanks exploded, and the sinking ship became a raging furnace. Two boats were lowered, but of the fate of the second the narrator had no knowledge. He remembered pulling desperately at an oar until the smoke cloud overwhelmed the boat. Then, gasping frantically for breath, he lost consciousness until he found himself on board the *West Barbican*.

At eight bells (8 a.m.) Peter was roused from his slumbers. A glance through the now open scuttle showed him that the ship was berthed alongside a wharf, and that the stevedores were already getting busy. A huge crane was transporting long, timber-protected pieces of steelwork into the *West Barbican's* No. 1 hold.

Peter regarded the steelwork with interest. It was the material on which rested the reputation and success of the Brocklington Ironworks Company, of which his father was managing director.

But other matters quickly demanded his attention. There was the damaged aerial. That had to be replaced under the direction of the Acting Chief Officer, but upon Mostyn's shoulders depended the responsibility of the perfect insulating of the wires. Already the necessary material had been "marked off", and the serang and his party were engaged in making eye-splices in the wire rope. At the mast-head of both fore and main, men were reeving fresh halliards for the purpose of sending the aerials aloft.

Captain Bullock was standing on the bridge watching the cargo being shipped, when he caught sight of the Wireless Officer. He beckoned Peter to approach. The officer of the watch was at the other side of the bridge superintending the securing of an additional spring; otherwise the bridge was deserted.

"Mr. Mostyn," began the Old Man abruptly, "I want you to understand clearly that there is only one captain on board this hooker, and he alone gives permission for officers to leave the ship. Who, might I ask, ordered you away in the lifeboat last night?"

"No one, sir," replied Peter.

"Then please remember that in future you are not to act on your own initiative except in matters directly concerning your duties as Wireless Officer. You were guilty of a grave breach of discipline. Don't let it occur again."

Mostyn smarted under this unexpected rap over the knuckles. He realized upon consideration that the rebuke was well merited. His offence was a technical breach of discipline. It was of no use telling this bluff old skipper his reasons. Yarns about "impulses of the moment" would elicit little sympathy. So he kept silent.

"All the same," continued the Old Man, in a less gruff tone, "you did a smart bit of work last night. Where did you learn to handle a boat?"

Mostyn flushed with pleasure.

"I've had three years in the Merchant Service, sir, and I've been in yachts and sailing dinghies ever since I can remember."

"I knew you didn't learn seamanship as a wireless man," continued the skipper. "Sorry I had to tick you off, my lad, but I simply had to. I'd like to send in a recommendation on your behalf, but I don't see how I can. Your Company would kick up the deuce of a shine if they knew I employed a wireless officer on executive duties. It's not done; or it's not supposed to be done—put it that way. And another thing: supposing, and it was quite likely, you'd lost the number of your mess over that business, what sort of yarn could I have pitched into the Board of Trade people? And my employers too? A pretty fine skipper they'd think I was, allowing a wireless officer to take away a lifeboat. Likely as not I'd have got the push from the Company's service and lost my ticket into the bargain. D'ye see my point?"

"Yes, sir."

"Then we'll cry quits. All the same it was a smart bit of work—a jolly smart bit of work—but I'll have to make an entry in the log recording the fact that you've been reprimanded and stating the reason. I don't think it will adversely affect you, Mr. Mostyn; rather the other way, I fancy."

Peter thanked the Captain and went about his duties, reflecting that the Old Man wasn't at all a bad sort, and that his bark was certainly worse than his bite.

Looking more like a blacksmith than a radio-operator, Peter completed his part of the work and applied the necessary tests. Everything was apparently in order in the wireless-cabin. With a grunt of satisfaction he replaced the receivers and left the cabin. Until the ship sailed—she was due to leave at ten that evening—he was at leisure.

"Now for a bath, a shave, and a change," he soliloquized. "It would never do to meet the pater in this state."

Somewhat to his surprise he found his father waiting in his son's cabin.

"Hello, Peter, my boy," was Captain Mostyn's greeting; "been ratting—or sweeping flues?"

Peter certainly looked a bit of a wreck. His sleepless night, following the perilous affair in the lifeboat, had given him a washed-out appearance. He was dog-tired, physically and mentally. He was dirty, unshaven, and rigged out in a very old uniform, with a scarf knotted round his neck in place of the regulation collar and tie.

"No, Pater," replied Peter. "Neither ratting nor sweeping flues. I've been choked off by the skipper."

"Easy job, judging by that running noose on your neck-gear," commented Captain Mostyn jocularly. "What's happened?"

Peter told him, simply and straightforwardly. There was never a lack of confidence between father and son. His parent listened attentively to the bald narrative.

"Your skipper was quite right," he observed. "In my days in the Service I wouldn't have thought of allowing a watch-keeping sub to go down to the engine-room and play about with the gadgets in order to slow down the ship. You did much the same sort of thing, chipping into a department that wasn't yours. At the same time, I'm proud of you, Peter. It shows you are not deficient in pluck. Right-o! carry on with your ablutions. I want to have a few words with Captain Bullock about the steelwork. While I'm about it I'll ask him to let you go ashore to lunch with me."

Captain Antonius Bullock was rather astonished to find that the managing director of the firm that had virtually chartered the *West Barbican* for three days was the father of his Wireless Officer.

"And I had to log him this morning," declared the Old Man.

"Yes, he told me about it," rejoined Captain Mostyn. "No, he didn't grouse about it. He quite sees the force of your argument. In fact, I told him practically the same thing."

"All the same," said Captain Bullock, "it was a smart piece of work. At my age I'd think twice before taking on a job of that sort. If I had to do it I'd do it, you'll understand, but these youngsters often rush into danger when there's no particular call for it; not their duty, in a manner of speaking. I'm rather curious to know what he did when that pirate collared the *Donibristle*. He told a lot about the affair, but precious little about his share in it."

"Peter had a pretty stiff time, judging from what he told me," observed Captain Mostyn. "Amongst other things he still bears the scars of eighteen wounds he received when the *Donibristle's* wireless-cabin was demolished by a shell."

"Eighteen, by Jove!" exclaimed Captain Bullock. "I had one—a beauty—in the war. Splinter from a four-inch shell when Fritz torpedoed the old *Harkaway* and fired on the boats. But eighteen!"

"Yes," commented Captain Mostyn. "He's seen more adventures during his short time in the Merchant Service than I did in thirty-seven years in the navy. During the whole of my sea service I never saw a shot fired in anger. Very good, I'll be on board at four o'clock to sign those papers. Do you mind giving my boy leave till then?"

Captain Bullock readily gave the required permission, and father and son had an enjoyable spell ashore.

By four o'clock most of the steelwork was safely stowed in the hold. Only a few crates of small parts remained to complete the all-important consignment for the Kilba Protectorate Government.

"That's all shipshape and Bristol-fashion, sir," remarked Captain Bullock, as the necessary signatures were appended to the papers in connection with the shipment. "If that precious lot isn't delivered safe and sound in Pangawani Harbour by the first of February it won't be the fault of Antonius Bullock."

CHAPTER VIII
The Passengers

At high water that night the S.S. *West Barbican*, drawing eighteen feet for'ard and twenty-four aft, left Brocklington Harbour, crossing the bar with less than five feet of water under her keel.

Fortunately the weather had moderated, the wind flying round off the land, otherwise she might have been detained for days, owing to the condition of the bar. The ship was now making for Gravesend to pick up passengers and mails, and thence for East Africa according to her usual programme.

Peter went on watch at ten that night with the unalluring prospect of remaining on duty till midday—perhaps longer—since Partridge and Plover, who had bucked up considerably during the vessel's stay in port, promptly showed signs of internal troubles the moment the bar was crossed.

It was not a prearranged case of malingering. There was no doubt about it: they had been ill. Neither knew of the burning of the oil-tanker, and of the dangerous position of the *West Barbican* when she proceeded to the rescue, until late on the following morning, and even then they received the news apathetically.

So Mostyn just carried on, pondering over the Company's doubtful economy, since, in addition to his normal pay, he was already raking in a fair sum for overtime in excess of the Merchant Service eight hours per day.

Gravesend was in its wonted late autumn state when the *West Barbican* dropped anchor. A thick fog entirely blotted out the shore. The air reverberated with the dismal hooting of sirens in every imaginable key; while bells clanging from vessels at anchor added to the din. At intervals the sun shone feebly through the yellow pall, although it was impossible to see twenty feet along the deck. To add to the general discomfort a raw, moist, west wind was blowing down London River, without having sufficient force to disperse the baffling fog.

The *West Barbican* was two and a half hours late in arriving at Gravesend. If she were to weigh at the scheduled hour the passengers would have to be smart in getting on board with their personal cabin effects. Their heavy baggage had been sent down to the docks and placed in a hold a week previously.

Peter Mostyn had turned in directly the ship dropped anchor. There was a chance of two hours well-earned rest, if rest it could be called, since he lay down on his bunk fully clothed save for his rubber deck-boots. It was one of those frequent occasions when he could not afford to waste precious minutes in dressing and undressing. He was almost too dog-tired to kick off his boots. He was dimly conscious of throwing himself on his bunk and pulling the collar of his greatcoat up over the back of his neck; then he passed into a state of oblivion, notwithstanding the discordant sonata within and without the ship.

He was awakened by the appearance of Mahmed with the inevitable char. The native boy was now in "full rig", a concession to the still-absent passengers. He wore a white drill suit, similar to that worn by officers in tropical climes, with the exception that there were no shoulder-straps. On his head he sported a round skull-cap of astrakhan, with a scarlet top.

"No come yet, sahib," announced Mahmed, in response to Peter's inquiry as to whether the tender had come alongside with the passengers.

"All right," rejoined Peter, as he handed back the empty cup. "Tell Partridge Sahib and Plover Sahib I want them in the wireless-cabin."

Going on deck, Peter found that the fog was as thick as ever. It was now nearly eight bells (4 p.m.), and the crew had been mustered for inspection. All the deck hands were now rigged out in uniforms. Instead of the motley garb, each man had a loose-fitting coat of butcher-blue, reaching to his knees and secured round the waist with a red scarf. His headdress was a scarlet, close-fitting cap, not unlike the Egyptian "tarboosh". This was the uniform issued by the Company for "ceremonial", and the expected advent of passengers was a fitting occasion for the display.

Three short blasts close alongside brought the officer of the watch to the end of the bridge.

"Tender alongside, sir," he announced.

The Old Man, in his best uniform, loomed up through the fog, disappearing as he hastened to the gangway, where, at the foot of the accommodation ladder, two lascars were stationed at the manropes to assist in the trans-embarkation of the passengers.

Gliding through the mist like a wraith the squat, snub-nosed tender ran alongside and was made fast. One by one the passengers began to ascend the swaying accommodation ladder. In all they numbered forty-one, mostly of the male sex. A few were missionaries bound for Kenya and Uganda; there were men taking up farming in the rich lands of the interior of British East Africa; mining engineers for Rhodesia; and people who for various reasons had booked their passages to the Cape by the *West Barbican* rather than by the fast mail-boats. There was also a young man in the uniform of a Mercantile Marine Officer. He was the ship's doctor, "signed on" for the voyage only, thus combining business with pleasure, being in ordinary conditions a hard-worked country practitioner. It was the first long holiday he had had for five years, and he meant to make the best of every minute of it.

There were seven lady passengers. The first one up the ladder was a stout, middle-aged woman, dressed rather startlingly for a trip on a tender in a fog. Her travelling-costume was certainly of good material but too vivid in colour for a woman of her age and build.

Mostyn, standing a few feet from the head of the accommodation ladder, watched her curiously. At one time she might have been good-looking. A perpetual sneer was on her face. She looked a woman who was habitually peevish and vile-tempered. Even as she came up the ladder she was complaining in a loud, high-pitched voice to someone following her — her husband apparently.

"Bet she's a tartar," thought Peter, and turned his attention to the next newcomer — a red-faced, sheepish-looking man, who, judging by his obvious bewilderment, had set foot for the first time upon a craft larger than a coastal pleasure steamer. Mostyn put him down as a country innkeeper, since he bore a strong resemblance to the host of the "Blue Cow" at Trentham Regis.

After that the crowd on the gangway thickened, the swaying ladder creaking and groaning under the weight of this queue of humanity. There were old men, young men; prosperous-looking men, poor-looking men; men with jovial lightheartedness written large upon their faces; others looking woebegone and dejected, as if regretting the past and dreading the future. There were men who might have been chosen as models in the rôle of Adonis; others who outvied in features the deepest Adelphi villain. Amongst the last of the arriving passengers came a girl of about nineteen or twenty.

She was slim and *petite*. Although wearing a serviceable raincoat she carried herself gracefully, holding but lightly to the handrail of the ladder. Mostyn noticed that her moist hair was of a rich, brownish hue, her features finely modelled. Her eyes were of a deep grey hue, beneath a pair of evenly arched eyebrows.

In spite of the clammy fog her cheeks shone with the glow of youth—a healthy glow that told unfailingly of an active, outdoor life.

"Jolly pretty girl, that," commented Peter, communing with his own thoughts.

The very last passenger to come over the side—Peter paid no attention to him—was a young, athletic man carrying a travel-worn leather portmanteau. With the air of one accustomed to life on shipboard he stepped briskly off the end of the gangplank and made straight for the saloon.

On the passenger list he appeared as William Porter, of Durban. Not one of the *West Barbican's* officers realized what viper the good ship was cherishing in her bosom; for in Berlin William Porter would have answered readily and truthfully to the name of Ludwig Schoeffer.

CHAPTER IX
A Quiet Trick

Some of the incidents in this chapter are based upon actual facts recorded in *The Signal*. The author takes this opportunity to express his thanks to the editor of that journal for permission, readily granted, to make use of certain incidents here recorded.

Mostyn made his way to the wireless-cabin to find his two satellites standing by according to orders.

"Well, all right now?" asked Peter solicitously.

"Yes, sir," was the reply in unison.

"What did you have for dinner in your mess?" pursued Mostyn, addressing Partridge.

"B'iled mutton, sir; and it weren't 'arf good."

"Not 'arf," corroborated the other bird. "An' b'iled peas an' dumplin's an' orl that."

"Right-o!" rejoined Peter briskly. "That shows you're both as fit as fiddles. We start sea routine at 10 p.m. You'll take on till four bells, Partridge——"

"Say, wot about my dinner?" objected the Watcher.

"Dinner?" repeated Mostyn, failing to grasp the reason of his subordinate's objection. "What's that got to do with it?"

"Dinner's at two bells, sir."

The Wireless Officer suppressed a desire to laugh.

"Four bells in the middle watch," explained Peter.

"That's 2 a.m. Surely to goodness you didn't expect to do a fourteen hours' trick? Plover, you relieve Partridge at four bells and carry on till I take over at eight bells—that's eight o'clock in the morning, not noon or four in the afternoon," he added caustically. "Got that?"

Yes, Messrs. Partridge and Plover had got that part all right.

"Now," continued Peter, "you know your duties. On no account touch the transmitter. Call me if there's any real need for it; and, don't forget, if you fall asleep on watch there'll be trouble."

Mostyn dismissed his assistants and donned the telephones. The *West Barbican* had weighed and was creeping cautiously down London River, over which the fog still hung as thickly as ever.

He anticipated a busy time. There were sure to be passengers who wanted to send messages at belated hours; urgent radiograms from shore stations, and radiograms that weren't urgent, were bound to be coming in; while, in addition, he had to deal with calls from ships and stations in the vicinity, and look out for time signals, weather reports, and possibly SOS and TTT warnings. Otherwise, save on approaching or departing from a port, the operator's work is light and at sea often approaching boredom.

Ten p.m. found the *West Barbican* rounding the North Foreland. She had now increased speed to nine knots, the weather becoming clearer. Hitherto, her passage down the river as far as the Edinburgh Lightship had been perforce at a painful crawl of four to five knots, with her siren blaring incessantly.

Mostyn had seen nothing of the passengers after their arrival. Being on duty he had missed dinner in the saloon. Not that he had missed much from a spectacular point of view, for most of the passengers were absent from that meal. A good many, in fact, would fail to put in an appearance at meals for several days, giving the hard-worked stewards and stewardesses a strenuous time in consequence. The latter were at it already, judging by the frequent popping of soda-water-bottle corks and cries of varying intensity and vehemence for "steward".

The tindal had gone for'ard and rung four bells. Peter, with the telephones still on, waited for his relief. Five minutes passed. He was beginning to think that the bird had played him false again, when Master Partridge's hobnailed boots were heard clattering on the brass-treaded ladder.

"Quite ready, boss," he observed genially.

Mostyn, without a word, handed him the telephones, repressing the desire to tick him off for unpunctuality. Then, waiting until the Watcher had adjusted the ear-pieces to his broad head, he wished Partridge "good night".

"Shall I turn in all standing?" he asked himself, as he switched on the light and surveyed his bunk. It was a bitterly cold night, for, with the partial dispersal of the fog, a cold nor'easter had sprung up. "A hundred to one I'll be routed out. Thank goodness we'll soon be in the Tropics!"

It did not take Peter long to turn in. For some minutes he lay awake thinking. He was far from easy in his mind concerning the Watcher on duty. In a congested waterway like the Straits of Dover and the English Channel—particularly in the vicinity of the Downs and off St. Catherine's— wireless messages of great importance to the safety of the ship and her passengers and crew might be sent; but would Partridge be alert enough to warn the *West Barbican's* operator? Supposing the bird fell asleep on watch? It was all very well for Mostyn to say that if a disaster should occur it would be put down to the fault of the system. That was not good enough for a conscientious fellow like Peter.

He resolved, in spite of his weariness, to make periodical visits to the wireless-cabin.

At 10.30 p.m. he cautiously approached the cabin; not with the idea of eavesdropping but merely to see if Watcher Partridge were on the alert. If he were, Peter meant to withdraw without disturbing him. If he were not— Peter smiled grimly.

Thrusting his feet into his rubber boots (on principle Mostyn always had sea-boots a size larger than he wore with shore-going kit) the Wireless Officer made his way to the cabin. A glance through the closed scuttle showed him that Partridge was wide awake, and that he still wore the telephones. Satisfied, he began to retrace his steps and encountered Preston tracking along the alleyway.

Dick Preston was still Acting Chief, the Chief Officer having failed to join the ship at Gravesend. Consequently the *West Barbican* was one executive officer short.

"Hello there!" exclaimed Preston. "Thought it was your watch below, Sparks. What's up: developed insomnia?"

Mostyn told him the reason for his visit to the bridge.

"That's all right, young fellah-me-lad," declared the Acting Chief. "You turn in. I know you've had a pretty sticky time. I'll keep an eye on yon greenhorn and see that he doesn't drop asleep on his perch. Trust me for that."

Five minutes later Peter was sound asleep.

Suddenly he was aroused by a hand grasping his shoulder. Only half awake the Wireless Officer sat up in his bunk, narrowly avoiding collision with the cork-cemented beam overhead.

"TTT, sir!" bellowed an excited voice.

For the present Peter was still hovering on the border-line 'twixt slumber and wakefulness. Somehow he had the idea in his brain that he was once more on board the S.S. *Donibristle*, and the officers' steward had brought him a cup of tea before going on watch.

"No, dash it all!" he expostulated. "I don't want tea now."

"TTT, sir! TTT!" repeated the disturber of Mostyn's peace.

Then Peter realized the situation. It was Watcher Partridge, almost falling over himself in his anxiety to proclaim the fact that at last he had had a call through of an important nature.

Tumbling out of his bunk, Peter slipped into his bridge coat, and hurried to the wireless-cabin, the Watcher, puffing and blowing, following hard on his heels.

Picking up the 'phones, Mostyn listened for a few seconds. Then he replaced the ear-pieces on the table.

"You'll have to do better than that next time," he observed caustically. "That's not TTT—nothing like it. It's North Foreland on our starboard quarter calling CQ. Tuning in, most likely."

Returning to his bunk, Peter noticed that it was now 11.15 p.m. There was still a chance of a good night's rest, he reflected.

At a quarter to twelve he was called again to receive time signals. Forty-five minutes later he was aroused to call for wireless orders for the ship. On this occasion nothing was forthcoming, so back along the now familiar alleyway he hurried to his sleeping-cabin.

It seemed as if Peter had been asleep only a few minutes when there was a terrific hammering at his door. Sitting up, Mostyn felt for the electric light switch. He found it easily enough. There was a metallic snap—but the cabin was not flooded with light. Something had gone wrong with the bulb, he reflected, as he shouted to the disturber without to come in.

The door opened. There appeared the perspiring face of Crawford, the engineer of the watch, his features thrown into weird relief by the guttering gleam of an oil hand-lamp.

"Hey, laddie!" he exclaimed in sepulchral tones. "Yon Watcher, he's——"

Words failed the Second Engineer.

"I'm awa' to sort yon," he added, and, as if no further explanation were necessary, bolted precipitately.

Imagining that nothing short of a vision of Partridge grilling on the main switch would meet his gaze, Peter doubled to the wireless-cabin. The alleyway was in pitch darkness. He collided violently with the Third Engineer, who, summoned from his slumbers, was making tracks for the engine-room.

On the bridge the officer of the watch was shouting to the serang to bring up the emergency oil-lamps. Every fuse in the ship had been blown out, and consequently not only the internal lighting had failed but the electric masthead and side lights had refused duty. With the *West Barbican* proceeding down Channel at fifteen knots on a dark night the possibilities of a disastrous collision were great, until the emergency lights were rigged up and the ship brought back on her course, since the binnacle lamp had failed with the other electric lights.

A strong smell of burning gutta-percha and ebonite greeted Peter as he gained the vicinity of the wireless-cabin. Outside stood Partridge and Plover, the latter about to take over the watch. Both were horribly scared, and no wonder, for upon striking a match Mostyn found the reason for all the trouble.

Watcher Partridge, on turning over to his opposite number, had hung the telephones on the main switch. He was deeply surprised and not a little pained when there was a miniature Brocks' display inside the cabin, both ear-pieces of the 'phones burning out and emitting most nauseating fumes, while every fuse on board had been blown out, causing a complete breakdown of the electric-light system.

After explaining matters to the angry Old Man, who was, figuratively, hunting for the scalp of the luckless Partridge, Mostyn set to work to rectify the share of the damage that came within his province. It took him the best part of an hour to replace the defective main switch by a new one, connect new telephones, and overhaul the set.

Then, back once more to his bunk, Peter realized that less than five hours remained before he took over the watch. It was now 3.15 p.m.

At 4.45 the engineer of the watch interrupted Mostyn's dreams. Once again the fuses had blown out, the cause being traced to the wireless-cabin.

The Wireless Officer stumbled across Master Plover at the foot of the bridge ladder. The Watcher was nursing his foot, and making inarticulate noises that denoted pain. The sole of his left boot was missing, together with the fearsome array of hobnails that used to play a tattoo upon the brass treads of the ladders.

Master Plover could give no coherent account of what had happened.

"I was sittin' there as quiet as a mouse a-listenin' in," he whimpered, "when I found myself chucked orf me chair right through the blinkin' door. S'elp me, I didn't do nothin' to the gadgets."

Peter guessed rightly as to what had actually happened. The Watcher wasn't watching. In other words, he had been dozing, and in a somnolent state had unconsciously placed his iron-shod boot upon the long-suffering main switch.

Making good defects, Mostyn managed to soothe the still highly nervous Plover into a state of tractability. Till a quarter to eight the jaded Wireless Officer did enjoy an uninterrupted sleep, then to be awakened by Mahmed's cheerful announcement: "Char, sahib."

Ten minutes later Peter took on. As he heard the dot-and-carry-one patter of the relieved Watcher's solitary boot, he smiled to himself and reflected that, although the work of a wireless officer is at times a strenuous one, it has its humorous side and is not without compensations.

CHAPTER X
The Unheeded SOS

During the rest of the day the *West Barbican* rolled before the following wind, to the no small discomfort of the majority of the passengers. It was a cold wind, too, and few of the passengers who had withstood the attacks of *mal de mer* ventured on deck.

"Have you found out who that loud-voiced female passenger is?" inquired Peter of Anstey, as the two paced the almost deserted boat-deck.

He put the question with ulterior motives, masking the main point of his curiosity.

"That queer specimen?" rejoined the Third Officer. "No, I haven't, beyond the fact that she's a Mrs. Shallop, and her husband, that red-faced man, is a horse-dealer, who made a pile in the war by stopping at home and selling broken-down hacks to Government inspectors who hardly knew the bow of a gee-gee from the stern. Yes, we're going to have some fun out of Mrs. Shallop before long, old son. She's had a row with the purser, two with the chief stewardess, and a few with the stewards thrown in as make-weights."

"What about?' asked Mostyn.

"Goodness knows," replied Anstey. "The purser was talking to the Old Man about it after breakfast. She's rather got on the poor chap's nerves. Apparently she's an imaginary grievance that they don't treat her like a 'lydy', so she's been ramming it down their throats that she's a naval officer's daughter—a captain's daughter."

"Well, isn't she?" asked Peter.

The Third Officer sniffed scornfully. Evidently Mrs. Shallop had fallen foul of him already.

"Naval captain's daughter!" he exclaimed. "Might be. Sub-lieutenants become captains, or at least some of them do; and subs have been known to do rash acts when they are young. But when a woman, whose accent, manners, and grammar are decidedly rocky, goes out of her way to assert that she's a naval officer's daughter, well then, snap goes the last thread of your credulity. My dear old thing, we're going to have some fun this trip, so get busy."

"Who is the girl—the girl who was almost the last on board?" asked Mostyn, broaching the long-deferred question at last. "Has she no friends on the ship?"

"Goodness only knows!" ejaculated the Third Officer fervently. "She's a Miss Baird, and I think she's by herself. We'll find out in due course. Hark! Yes, at it again! It's poor old Selwyn getting it this time."

Through a partly open skylight came the now familiar voice of Mrs. Shallop, almost ear-piercing in its intensity and raucous in its tone. Mingled with the strident outbursts of the woman came short, incompleted protests from the doctor, who apparently was not able to hold his own.

"At it again," reiterated Anstey. "She's trying the naval captain stunt on the doc. I guess—by Jove! Wait till she tackles the Old Man."

Just then Dr. Selwyn appeared on the boat-deck. He was a dapper little man with the reputation of being a skilful and rapid surgeon. He could have commanded a large practice in town, but, preferring the country to city life, was content with a moderate income and plenty of hard work in congenial surroundings. In manner he was affable, and possessed an old-world courtesy that made him extremely popular. He was mild in speech, and rarely lost his temper; but when he came on deck it was obvious to both Peter and Anstey that he was labouring under suppressed anger.

"Morning, Doc," was the Third Officer's greeting. "Up for a breather?"

Selwyn braced his shoulders and gazed out to starboard. Nine miles to the nor'ard the white cliffs of the Isle of Wight stood out clearly against the dark grey clouds.

"Yes," he agreed. "A breather. Had a fairly stiff time with sundry patients. Sort of thing one must expect in the early days of a voyage. What's that land over there?"

"St. Catherine's," replied Anstey. "If it's clear enough we may sight the Isle of Purbeck, but I doubt it. So take your last look at Old England for a while, Doctor."

The three men remained in conversation for several minutes, but Anstey failed hopelessly in his attempt to "draw" Selwyn with reference to his encounter with the "tartar".

"I'd like to see your wireless-cabin," remarked the doctor.

"Certainly," agreed Mostyn. "As a matter of fact I'm about to take over the watch."

Anstey, to whom the wireless-room was no novelty, "sheered off" and shaped a course for the smoking-room, while Peter and the doctor made their way for'ard to the former's post of duty.

Suddenly Peter stopped. From the open door of the wireless-cabin came the deep bass voice of Captain Antonius Bullock. He was "letting rip" vigorously, and there was anger in his tone. Then, trembling like a leaf, Watcher Plover appeared.

The Old Man, paying an unexpected visit, had found the Watcher fast asleep.

Already the skipper was "fed up to the back teeth" (to use his own words) with the two birds. Coming on top of the disconcerting incidents of the night, when both Watchers had severally dislocated the electric-lighting service, Plover's delinquency, serious enough in any circumstances, completely upset the Old Man's equilibrium.

By this time he was fully convinced that the Watcher system was rotten to the core. On his previous voyage Captain Bullock had fallen foul of his wireless officers, but that was over technical matters. Otherwise he had had no cause for complaint, and, generally speaking, the relations between skipper and radiographers were harmonious if not exactly cordial. Now, thanks to a misguided attempt at economy, the Old Man could put no dependence upon Mostyn's assistants, and, in fact, he was inclined to blame Peter for not exercising more supervision over his subordinates.

Which was rough on Peter. In Captain Bullock's present mood it was useless to point out how many times during his "watch below" Mostyn had been called to the wireless-cabin. The fact remained that Partridge and Plover had been signed on for the trip. Even if the Old Man wished he could not land them this side of Las Palmas, and so for the present Peter must make the best of things, trusting that in due course the two incompetents might be "licked into shape".

As soon as Captain Bullock had retired to his cabin, Peter took over the watch, Selwyn standing by as the Wireless Officer made the usual tests.

"Now you can listen in, Doctor," announced Mostyn, after he had produced and connected up a supplementary pair of 'phones. "There's not much doing, I fancy."

Selwyn adjusted the ear-pieces, while Peter, similarly equipped, stood by pencil in hand in order to give his companion some inkling of any stray message.

"There's something!" exclaimed the doctor. He was excited. As cool as the proverbial cucumber when he was performing a deft and rapid operation upon which human life depended, he was now as delighted as a child with a new toy, when he heard the high-pitched buzzing sound that indicated a message in transit.

"Niton," explained Peter. "Isle of Wight station. She's calling up—no—half a minute."

Mostyn's pencil was moving rapidly as he recorded the message.

"Cut out o.m. SOS signals out: stop sending."

Then almost immediately after came a plaintive wail from a ship:

"Please repeat whole of preamble and words after 'overcoat'."

"Explain, please," asked Selwyn.

Mostyn, busy altering the wave length in an attempt to pick up the SOS, did not reply. Explanations could come later.

A vessel fifty miles away was trying to obtain a repetition of a message from Niton. Part of it she had received, but her operator was doubtful about the preamble and the words following overcoat. It was a purely private message, of no interest to anyone save the sender. Niton was trying to make the operator stop sending, as there was an SOS message coming from somewhere. The ship's operator for some reason was persisting in his inquiries for the words following overcoat. In addition a distant high-power station was chipping in, and there were also "atmospherics" of high frequency.

Out of this chaotic "jam" Mostyn was trying to isolate the urgent wireless call for aid.

Almost deafened by the exaggerated reverberations of the ear-pieces as Mostyn pursued his efforts to tune in, Selwyn watched with unabated interest the Wireless Officer's deft manipulations of the set. Greek the doctor understood, but this was something far beyond his ken.

At last. Faintly, almost indistinguishable from the cackling of the atmospherics, came the despairing SOS. It emanated from a vessel in dire distress. Peter knew that she was using her comparatively low emergency set. That indicated the fact that her ordinary sending apparatus had broken down.

"SOS. S.S. *Passionflower* 17 miles s. by w. of Owers. Boiler explosion, ship making water rapidly; pumps inadequate."

"Message received," sent Mostyn, then handing Selwyn the paper on which he had written the fateful message, "Captain, please," he said.

The doctor removed his telephones and departed on his errand. Meanwhile Mostyn was listening in for other vessels in the vicinity replying to the general and urgent call for aid.

In the chartroom the Old Man and Preston held a hasty conference. Only an hour previously the *West Barbican* must have crossed the track of the disabled *Passionflower*, within a few miles of her. Now a distance of between fifteen and twenty miles separated the two vessels, and to render assistance the former vessel would have to retrace her course. At fifteen or fifteen and a half knots it would take her more than an hour to close with the *Passionflower*. If she did, would she be the first on the scene?

Both the Old Man and the Acting Chief Officer doubted it. This part of the Channel was a busy one. Not only was there the "up and down" traffic, but a large number of vessels was plying between Southampton and the Normandy ports. In addition, the *Passionflower* was within an hour's run of Portsmouth, where there were Government tugs and destroyers ready to render aid.

The navigating officer's doubts were confirmed when Mostyn appeared with a report that already five vessels were proceeding to the rescue of the *Passionflower*. So the *West Barbican* held on her course.

A little later Peter, who had contrived to "cut out" the plaintive and persistent inquiry as to the words following overcoat, got into touch with the P. & O. liner *Nowabunda*. From her he learnt that the *Passionflower* had been sending out her SOS for an hour before the *West Barbican* had picked up the distress call.

Either Watcher Plover had been asleep for some time before being awakened by the skipper, or else his untrained ear had failed to detect the low notes of the distressed vessel's emergency set. The actual result was the same. The *West Barbican*, although nearest to the *Passionflower* when she first began the call for aid, had passed by unheedingly. Had she proceeded to the spot she could have towed the crippled vessel into Portsmouth or Southampton with very little difficulty.

This is what the Portsmouth tug *Sampson* did, the *Passionflower* being dry-docked just in time to save her from foundering. In the Admiralty courts the salvage earned the *Sampson* £11,000, and this the *West Barbican* lost simply and solely through Watcher Plover's incapacity.

CHAPTER XI
The Old Man is Disturbed

Captain Antonius Bullock had turned in for the night. He had received the reports of the officer of the watch and the engineer of the watch, the time signals and weather reports from the Wireless Officer, and was now free from the cares of command until such time as his steward called him. He might be called within the next minute; but with luck he hoped to remain undisturbed until six bells in the morning watch.

It was now 1 a.m. The *West Barbican* had passed Ushant twenty miles to port, and was entering the Bay of Biscay.

The weather was still cold, but the wind had moderated considerably, coming off the land. The Bay was on its best behaviour, and consequently the passengers, who were beginning to find their sea-legs, were wandering farther afield than the limited expanse between the saloon and their respective cabins.

The notice on the Old Man's door, "Don't knock, come in", had disappeared. Captain Bullock had seen to that. It served its purpose when the ship was getting ready for sea, but once the passengers came on board the brusque invitation vanished.

Although the air without was raw it was cosy and warm inside the cabin. The radiators, heated by steam from the boilers, kept the apartment at an even temperature, while, as a concession to appearances, a fire glowed in a polished, brass-mounted grate. Only no heat came from that fire: it was a dummy, composed of coloured paper rolled into loose balls and packed around an electric-light bulb. It had a comforting look, and frequently visitors to the Old Man's cabin stood on the hearthrug enjoying the heatless glow in utter ignorance of the fact that no fire burned in that polished brass grate.

Over the door and scuttles the dark-blue baize curtains had been drawn. The electric light had been switched off, and only the red glow from the grate faintly illuminated the cabin.

Captain Bullock lay in his bunk, raising his head occasionally to sip at a stiff glass of special Scotch. From early morn to midnight he was a rigid teetotaller Even at dinner the decanters passed by him untouched, but every night, even in the hottest weather, his steward mixed a uniformly strong glass of whisky, hot water, and lemon.

Generally the Old Man was quickly asleep, but to-night he felt wakeful. Not as a rule a deep thinker—he was essentially a man of action—he found himself pondering over various matters.

He was beginning to realize that this was his last voyage. On the *West Barbican's* return to London he was to relinquish his command and retire on pension. How he hated the idea! The sea was part of his being. No one knew the call of the deep more than he. True, at times, he had been "fed up" with the sea, but those were only passing moods. Some men looked forward to superannuation from the time they entered seriously into the battle of life. They had visions of peaceful if not luxurious retirement, living happily and contentedly on their hard-earned pensions. "And usually," thought Captain Bullock, "they are dead in a couple of years—rusted out through sheer idleness."

No, he hated the idea of having to "go on the beach" for the rest of his life. Settling down in the country and keeping fowls did not appeal to him in the slightest. He might get a job as harbour-master in some minor port, but these ports are limited in number. Besides, he did not take kindly to the idea of being badgered by a petty Harbour Board, the members of which were probably coal-dealers and corn-factors who knew nothing about the sea.

"Here I am, as hard as nails, sound as a bell, and a better skipper than I was twenty years ago," he soliloquized. "Why can't the Company keep masters on till they show signs of cracking? They'd get something for their money instead of paying it out in pensions."

Then his thoughts reverted to the lost opportunity of the *Passionflower* salvage job. True, there was the business of the oil-tanker *Bivalve* as a set-off, but he wondered what his owners would think when they read of the case in the *Shipping Gazette*.

Suddenly his reveries were interrupted by the sound of the cabin door lock being turned very cautiously. The sound was barely audible above the varied noises without.

By this time Captain Bullock was in a drowsy state. Without raising his head from the pillow, he was dimly aware that some one had entered the cabin. It was unusual. Sometimes his steward had occasion to enter during the night. Occasionally the officer of the watch or the Wireless Officer brought a report, and in any case they explained their presence verbally.

"Perhaps he thinks I am asleep and doesn't want to disturb me," thought the drowsy man, and, without attempting to fix the intruder's identity, he lay still, apathetically watching the other's movements.

The intruder crossed the cabin silently yet without hesitation. He stood at the writing-desk for a brief instant and then withdrew.

"'Spose it's Anstey with a chit," decided Captain Bullock, and, satisfied with his own explanation, he fell asleep.

At 6 a.m. the Chief Steward mustered his staff preparatory to the usual routine. There was an absentee: the Captain's steward.

"Anyone seen Wilkins?" demanded the Chief Steward.

No one had. Some one dispassionately volunteered the information that Wilkin's bunk had not been slept in. Men roused from slumber to perform the irksome routine are apt to be apathetic before breakfast.

The Chief Steward dismissed his staff to their various duties, and proceeded to search for the missing man.

He found Wilkins fully dressed and fast asleep on the floor of the pantry. On a shelf stood an empty tumbler that smelt of whisky.

The Chief Steward stirred the sleeping man with his boot.

"Come along," he exclaimed. "Show a leg, there! Skipper's waiting to be called."

Beyond a protesting grunt Wilkins showed no sign of recognition.

"Drunk as a lord," commented the Chief Steward. "Come on, man!" he added sternly. "Pull yourself together. You've been after the Old Man's whisky-bottle."

A friendship existed between the two men. The Chief Steward had obtained Wilkins's post for him. In consequence the former made allowances, which he would not have done in the case of another of his subordinates.

Holding Wilkins under the arms the Chief Steward dragged him unceremoniously along the deserted alley-way, and bundled him into his own cabin. There he would be safe from detection.

Locking the door, the Chief Steward returned to the pantry, washed out the tell-tale tumbler, and then summoned an assistant steward.

"Wilkins is ill," he announced briefly. "Take on Captain's steward's duties until he's fit again."

At five minutes to seven Assistant Steward Scott, bearing a can of hot water and a cup of tea, tapped at the Old Man's cabin door.

Captain Bullock, as fresh as a proverbial daisy, eyed the deputy coldly. Any alteration of routine jarred him.

"Where's Wilkins?" he demanded.

"On the sick list, sir."

"Humph. Bath ready?"

"Yes, sir."

The Old Man donned his bridge coat over his pyjamas before making tracks for the bathroom.

Suddenly he turned to his servant:

"So you were the man who came into my cabin during the middle watch?"

Scott stammered and went very red in the face. He was a meek, inoffensive man, and stood in deep awe of those set in authority over him.

"No, sir. Please, sir, I didn't," he protested. "I only took on at four bells."

Captain Bullock made no audible comment. He went to the writing-desk to see if anyone had left a chit there. There was none. He gave a swift, comprehensive glance at the book-shelf where, among other volumes, were the three separate code-books by which the owners and consignors were able to communicate with the ship. They were in their usual places.

The Old Man smiled grimly as he put a hastily formed suspicion from his mind.

"All right," he said gruffly. "Carry on."

CHAPTER XII
The Code-book

Mr. William Porter—otherwise Ludwig Schoeffer, had taken readily to his new surroundings on board the S.S. *West Barbican*. He made it a habit to do so, wherever he was: at the Wilhelmstrasse, Berlin, or in Sing Sing Prison, New York. He made a speciality of studying men and things, and, in order to do so, he naturally came to close quarters with the objects of his professional attention.

He had failed to prevent the shipment of the Brocklington Company's consignment of steelwork for the Kilba Protectorate. There remained a chance of achieving his object while the steelwork was on the high seas; and to that end he had booked a passage in the *West Barbican*.

His primary idea was to sink the ship without loss of life. It might have been a new-born hesitation to take human life that actuated his plans. During the war he had not been so scrupulous. Now, perchance, he looked upon murder and manslaughter in a different light. Or perhaps he was developing nerves and was afraid of falling into the clutches of the law, for he knew full well that, if he bungled, his employers in Germany would utterly repudiate him.

It might have been possible for him to place a delayed-action infernal machine in the hold of the *West Barbican* when she was loading up at Brocklington. But he had not an intimate knowledge of the construction of the ship, and he feared to take drastic steps without being certain of his surroundings. Nor did he wish to immolate dozens of passengers.

The majority of the latter would be leaving the ship either at Cape Town or Durban, so their departure would ease the situation as far as the remnants of his conscience were concerned.

He decided, therefore, to go as far as South Africa as a passenger on the *West Barbican*. During the voyage he could obtain a good knowledge of the ship's routine, and the accessibility or otherwise of the holds and bunkers. Then, before leaving the ship at Durban, he could "plant" his high-explosive bomb and send the *West Barbican* to her doom.

It was an easy matter to convey the explosives on board. The customs officers at British ports are vigilant enough in connection with homeward-bound passengers' baggage, but not so in the case of departing ships. No one paid any attention to the dark-red, cloth-bound book that Mr. Porter carried under his arm. It never occurred to Ludwig Schoeffer that it was hardly fair to a book to be carried so openly on a damp, foggy day.

Outwardly it was a book, but between the covers there were no leaves except dummy edges. In the recess thus formed was four pounds of very high explosive, sufficient to blow a hole completely through the steel plating of a merchant-ship's hold. The explosive without a primer was comparatively innocuous. It could be subjected to a severe blow without detonating; fire had no effect upon it, except that it would smoulder without bursting into flame. But when mixed with a solution of potash the latent power was instantly and terrifically released.

Until the bomb was prepared for action Schoeffer kept the glass tube containing the potash separate from the main explosive. If necessary he could easily explain the potash by saying it was medicine.

The detonation of the infernal machine was actuated by a fairly simple device. It was only necessary to smash the glass tube of potash; but the point was: how could Schoeffer break the glass when he was away from the ship?

If anyone had had an opportunity of inspecting Mr. Porter's watch he would certainly have been interested; for, in addition to the hours, minutes, and seconds hands, the dial sported a hand that indicated the days up to seven. But in place of numbers on the day circle there were seven black dots. Each of these dots proved to be a small insulated metal peg, capable of being raised until it projected a fraction of an inch from the dial, yet sufficiently to hold up the hand.

To complete the outfit there was a small eight-volt battery, which, on a circuit being formed, would detonate a minute charge of explosive, enough to smash the glass tube, liberate the potash, and cause the desired catastrophe. By means of the watch Schoeffer could delay the explosion from one to seven days after he had set the bomb in position.

Mr. Porter made rapid strides in forming acquaintances on board. He was affable without being obtrusive; communicative up to a certain point, without volunteering information; a good conversationalist without boring his listeners. He took a keen interest in the officers, the stewards, and even the lascars, but, in the course of conversation with them, he rarely if ever asked questions concerning their professional duties.

One person in particular he cultivated. That was Wilkins, the Captain's steward. Wilkins was a professional postage-stamp agent; he bought large quantities of stamps in foreign parts on behalf of a London firm. Mr. Porter was a keen amateur collector, and so a bond of interest was formed.

Since the facilities for encouraging conversation between passengers and stewards are limited, Schoeffer found a convenient opportunity to confer with Wilkins on the subject of postage stamps. The opportunity occurred just before "lights out", the venue being the pantry.

Schoeffer found that the subject of stamps afforded him a splendid chance of gaining information concerning the Old Man. He knew that the skipper kept the code-books in his cabin. Two of them—the *ABC* and the *Telegraph Code*—were practically public property, but the third was the private code of the Blue Crescent Line, by which the owners telegraphed orders to their various ships.

The German agent made no attempt to suborn the steward to "borrow" the code-book. He preferred to work single-handed. It was infinitely safer. But he soon discovered that Captain Bullock was a light sleeper and that he was practically an abstainer from strong drink, except for his regular "night-cap".

One night the chance occurred. Wilkins had mixed the Old Man's grog. His attention diverted for a minute, he was unaware that Mr. Porter had dropped into the glass a cube resembling sugar but containing a powerful narcotic quite devoid of taste.

"Well, sir," remarked Wilkins, "I must push off and take this to the skipper."

With this gentle intimation the steward speeded his guest. He had reasons for so doing. He had no desire to let even an affable gentleman like Mr. Porter know that he was in the habit of helping himself to the Old Man's whisky.

A few minutes later Wilkins poured out another stiff glass of grog and carried it to the skipper, leaving for his own consumption the glass that Schoeffer had doped.

Ten minutes later the steward returned to the pantry, drunk the doctored whisky, and spent the rest of the night in a state of insensibility, in which condition he was found and befriended by the Chief Steward.

Returning to his cabin—a single-berth one on the port side—Schoeffer closed the deadlight and drew a curtain over the jalousied door. At twelve the electric lights in the passengers' cabins were switched off, but that hardly troubled "Mr. Porter". An electric torch gave him all the light he required.

Two bells sounded. Cautiously Schoeffer switched off the torch, undressed, and put on dark-coloured pyjamas and felt bedroom slippers. Then, after listening to hear that no one was about, he stole silently from his cabin.

He guessed that the officer of the watch would be drinking cocoa in the chartroom, and that the bridge would be deserted save for the native quartermaster at the wheel. If he were intercepted, Schoeffer would pose as a somnambulist and suffer himself to be led back to his cabin.

But no one was about. Boldly yet stealthily he gained the bridge and entered the skipper's cabin, confident that the Old Man was in a drugged sleep. He would have had a nasty shock had he known that Captain Bullock was merely drowsy and was aware of his presence.

With the private code-book in his possession Schoeffer retraced his way to his cabin. Luck was with him. Unseen and unheard he entered his stateroom and closed the door. For the next two hours he was hard at work carefully copying out cryptic letters, that in due course would enable him to carry out his nefarious plans to perfection. He also carefully committed to memory the instructions printed in the front of the book relating to the procedure to be followed in sending and receiving instructions by code.

Again he sallied forth to the Captain's cabin and replaced the book. What rather puzzled him was the fact that the Old Man was sleeping naturally. His deep, regular breathing did not conform to the suggestion that he was under the influence of a powerful drug.

It was a disquieting discovery. He could not account for it. Perhaps, he thought, Captain Bullock had something up his sleeve. Even the satisfaction of having secured and made full use of the secret code-book had much of its greatness shorn by the haunting dread of the burly captain of the S.S. *West Barbican.*

CHAPTER XIII
Crossing the Line

"Mr. Mostyn."

"Sir?"

"Did you by any chance use the owner's code-book during the middle watch?"

"No, sir."

"Very good; carry on."

This was the brief conversation between the Captain and the Wireless Officer. The Old Man had by some unaccountable intuition fostered the idea that the code-book was the object of the intruder's presence. Mostyn had a right to make use of it, and, before probing deeper into the problem, Captain Bullock had questioned him.

The skipper had a keen insight into human nature. In his official capacity he had come into contact with hundreds, nay thousands, of human beings for whose safety and welfare he, under Providence, was responsible. Some were notables, the majority common-place individuals, and not a few persons with unenviable reputations. He had had on board escaping murderers, defaulting company promoters, fraudulent trustees, absconding cashiers, and a variety of other criminals from the "flash" cracksman to the common "lag". Professional gamblers, sharpers, and pickpockets had passed his way on the broad highway between Great Britain and the Dominion of South Africa.

Captain Bullock was generally very quick in "knowing his man". Rarely was he mistaken in his speedy yet calculating judgment. Already he had his Wireless Officer "sized up", and the verdict was favourable. Hence Peter Mostyn's "No, sir," was sufficient. The Old Man knew that he had spoken the truth and that he was not the mysterious intruder.

Anstey, the officer of the watch, was likewise questioned. He, too, was emphatic that he had not entered the Captain's cabin, nor had he seen anyone doing so during the middle watch.

For some days Captain Bullock pondered over the incident, blaming himself for not having challenged the intruder. Then he began to let the matter dwindle in importance, and by the time the ship reached Las Palmas he had practically forgotten all about it.

In fine, excessively hot weather the *West Barbican* approached the Line. No tropical storm greeted her as she entered the once dreaded Doldrums, that belt of calms which has yielded its powers of holding ships captive for days on end, to the all-conquering steam and internal-combustion engines. Rarely now is there a sailing-ship to be sighted wallowing helplessly in the Doldrums, her decks and topsides opening with the terrific heat, and her crew driven almost mad with the torturing glare of the tropical sun. Auxiliary power has changed all that, and even the huge, square-rigged ship engaged in trading round the Horn is now equipped with a semi-Diesel capable of pushing her along at a modest four or five knots in a calm.

Preparations to pay the customary honours to Father Neptune were in full swing on board the *West Barbican*. For days before the ship was due to cross the Line all the officers and twenty-five per cent of the passengers became temporary inquiry agents. Seemingly casual conversation was entered into with the primary object of discovering who had or who had not "crossed the Line". Within a few minutes of an unguarded remark being made by a passenger to the effect that he had not been in southern latitudes, that fact was duly recorded in a notebook by the indefatigable Acting Chief Officer. Preston was a veritable sleuth-hound in these matters, and already his "bag" was assuming favourable proportions.

Among the names recorded were those of Partridge and Plover. The two Watchers had never heard of the time-honoured ceremony, and were in utter ignorance of the ordeal through which they would have to pass. Their lack of general knowledge, combined with a somewhat surly reticence, had made them no friends on board. They kept to themselves, hardly exchanging a word with anyone else except when duty compelled them to speak.

At length the eventful day arrived when the ship was due to cross the parallel of maximum length. Soon after day-break eager lascars had been employed in spreading a huge tarpaulin over a rectangular frame, so as to form a large bath. At one end, facing the for'ard portion of the promenade deck, a platform was erected and draped with bunting. Behind locked doors officers off duty lurked in their cabins, contriving weird and startling disguises for the Sea King's festival. The donkey-engines were started — not with the idea of ejecting bilge water, but for the purpose of pumping a copious supply of salt water into the improvised tank.

On the bridge Preston was "shooting the sun". Again and again he levelled his sextant, until he was satisfied that the ship was within a few miles of the Line. Then, hastily reporting the fact to the Old Man, he disappeared down the companion-ladder to change with the utmost speed into a wondrous garb comprised chiefly of a bathing-suit, seaweed, and oyster-shells. Next, assisted by an individual who resembled a cross between George Robey and Little Tich, and who was to appear as the doctor, Father Neptune donned flowing locks and beard of picked oakum, assumed a massive crown of tinsel, and grasped his trident.

At that moment the ship's siren gave a terrific blast. It was the signal that Neptune's cortège had been sighted by the look out for'ard.

The fo'c'sle and foremost shrouds were packed with eagerly gesticulating lascars; native firemen squatted on the decks on either side of the tank, and clung like flies to the stanchion-rails. On the promenade deck all available camp-chairs had been pressed into service and were occupied by excited passengers, trying to keep cool in vain, in spite of the double awnings.

Presently Captain Bullock, resplendent in white tropical uniform with gilt buttons and shoulder-straps, descended from the bridge and took up a position in the centre of the front row of crowded deck-chairs.

"Ahoy!" roared a deep voice for'ard. "What ship is that?"

"The S.S. *West Barbican*, of and from London," bawled the Old Man in reply.

"Then harkee, Skipper. Father Neptune demands entrance and the honour due to his exalted rank."

"Come aboard, sir," rejoined the Old Man.

Heralded by a fanfare from hand fog-horns, and a terrific din from a variety of metal implements, begged, borrowed, or stolen from the galley, Father Neptune appeared not exactly over but close to the bows. Brandishing his trident he bellowed a nautical greeting, and proceeded to assist his Queen through the limited space of the hatchway. It was soon evident that the lady was in difficulties and a plainly audible, "Steady on, old man," delivered in a very masculine voice, had the effect of raising a boisterous chorus of laughter from the sightseers.

Amphitrite, disentangled from the embraces of a catch on the hatch-cover, appeared in her lord's wake, but the effect of her flowing locks of golden hair and her deeply rouged face were somewhat marred by the display of a pair of unmistakably masculine hands and feet.

The doctor and the barber next struggled for publicity, each questioning the other's right of precedence, with the result that each contrived to get his head through the hatchway and no farther.

It was not until the barber had converted the doctor's hat into a concertina that the former contrived to make a complete appearance, followed by the doctor, who, in his broad Scotch that betrayed him as M'Turk the Chief Engineer, requested his companion "not to play the fule beforr your time".

Then came the bears—grotesquely garbed fellows recruited mainly from the Chief Steward's department, but with the residue of the engineers off duty to leaven the whole lump. Almost before King Neptune and his Queen were seated upon their respective thrones the zealous bears had scattered to rope in the victims of the revels.

The first to be brought into the arena was Watcher Partridge. His opposite number, scenting trouble, had deserted him, and was making his way to the stokehold, hotly pursued by a couple of brawny bears.

Partridge submitted sullenly. Without a word or act of protest he was led before the doctor.

"Are ye no' weel, laddie?" inquired the doctor. "Open your mouth and show your tongue."

The bird obeyed.

The next instant he was spluttering and coughing, for the doctor had dexterously placed a pill, composed of the unholiest ingredients of the engineers' stores, in the wide-open cavity. Still spluttering, he was again seized by the attendant bears, blindfolded, and forced into the barber's chair.

The barber eyed the agitated Partridge dispassionately.

"Hair cut or shave?" he inquired, and, receiving no reply, he seized one of his razors, a formidable-looking instrument fashioned out of a barrel stave.

A few deft strokes and the deed was done. Partridge, released from the chair, sprang to his feet amidst the delighted howls of the spectators. One side of his face was streaked with Stockholm tar, the other with red ochre.

"Run for it!" exclaimed one of the bears, guiding the bewildered Partridge towards the tank. The bird hopped it, trod on air as one foot overstepped the narrow edge, and, with a sousing splash, he plunged headlong into the water.

He had barely time to gasp for breath when a bear ducked him. Thrice this operation was repeated before the pie-bald Watcher was allowed to escape, without even receiving King Neptune's congratulations upon becoming a Son of the Sea.

The while other victims were being attended to by the doctor and the barber, and unceremoniously bundled into the tank.

For the most part they accepted the situation with a good grace. In the case of the passengers who had not crossed the Line before, certain allowances had been made for them; nevertheless some were rather rigorously handled before receiving their diplomas as Freemen of the Seas. Since they had received short notice to the effect that it would be as well if they "rigged out" to be in readiness for a ducking, they took the hint, changing into bathing-costumes or any old clothes obtainable.

One passenger, a burly, six-feet-two individual, with huge biceps showing up under the tight sleeves of his bathing-suit, certainly gave the bears a run for their money; for, when they went to bring him to Neptune's court, they found that he had put on a pair of boxing-gloves.

"Come on!" he exclaimed, with a good-tempered laugh. "I'll take on the whole crowd, Neptune included."

Nothing loth, a plucky little bear stooped and rushed in to collar the defiant passenger round the waist. The next instant he was sent staggering into the arms of one of his companions, and the two floundered on the deck, capsizing the barber and his two pots of ochre and tar.

"At him, lads!" roared Neptune, forgetting in his excitement that he was playing the rôle of King of the Sea.

Five or six bears rushed at the man from opposite sides. He waited until they were almost on him, then, without the faintest sign of his intention, dived straight at the feet of those on his right.

There was weight and power behind those hunched shoulders. Three of his assailants, swept off their feet, crashed to the deck, while their comrades, unable to check the impetus of their rush, tumbled in a confused heap upon the baffled, sprawling three.

From under this struggling mob, like a porpoise in an angry sea, emerged the stalwart passenger. Springing to his feet he dashed up the ladder to the promenade-deck, cleared a way between the throng of spectators, who cheered him heartily, and gained the boat-deck.

For a while he paused to contemplate the sorting out of the discomfited bears; then, finding his pursuers hard on his track, he scaled the side of the wireless-cabin. On the roof he took up his stand. With his broad back against the trunk of the aerial it looked as if he could hold his own against all comers.

The lascars were beside themselves with excitement. The passengers, leaving the shelter of the double awnings, stood under the blazing sun, straining their eyes in the dazzling glare as they watched the tactics of their champion.

"Lasso him, lads!" shouted Neptune, laying aside his trident and preparing to take an active part in the subjugation of his recalcitrant subject.

Some of the bears hurried off to obtain ropes. Others waited by the base of the wireless-cabin, feeling decidedly uncomfortable as the hot sun played upon their scanty, wet garments.

Just then another party of bears came for'ard dragging the luckless Plover, whom they had captured in an empty bunker.

The appearance of the second bird created a diversion. The bears guarding the wireless-cabin, eager to witness the initiation of the unpopular Plover, lost interest in the huge passenger on the roof.

In a trice the latter slid down to the bridge, swung himself down by a stanchion to the promenade-deck and thence to the enemies' camp—the temporary court of Father Neptune.

Hurling aside the doctor, who had already received rougher treatment than he had meted out to his victims, the defiant subject of King Neptune made a bull-like rush for that august monarch.

The next moment they were at grips. In spite of wearing boxing-gloves the stalwart passenger held Neptune tightly round the waist. The latter strove with his sinewy hands to disengage himself from the powerful embrace. In the struggle Neptune's tinsel crown slipped over one eye and his tow-beard fell off, revealing the rugged features of Acting Chief Officer Preston.

For about thirty seconds the two men struggled furiously, yet the keenest observer could detect no trace of bad temper. The adversaries were sportsmen both, who knew how to keep themselves under control.

With the sweat pouring in streams down their faces they continued swaying and heaving. Both were of about the same weight and build. Preston had the handicap of about ten years, but he was as fit as a fiddle and hard as nails.

Amphitrite had discreetly retired from the arena, while the bears, unwilling to take an unfair advantage of their intended prey, stood in a semicircle, impartially encouraging both adversaries. Even Captain Bullock, who through long usage had become bored stiff with the "crossing of the Line revels", was on his feet shouting excitedly at the novel spectacle of Neptune being bearded in his den.

Suddenly the unexpected climax happened.

Before anyone could utter a warning or check the impetuous movement of the two wrestlers, Preston was forced to the edge of the temporary dais, which was on a level with the wire guard-rails.

Probably his antagonist was blinded by the perspiration running into his eyes, because he failed to see the danger resulting from his headlong rush.

Locked in each other's arms the two men disappeared over the side of the ship.

**THE TWO MEN DISAPPEARED
OVER THE SIDE OF THE SHIP**

CHAPTER XIV
Mostyn to the Rescue

For a brief instant the danger and suddenness of the catastrophe were hardly realized. Assembled for a pageant the passengers were horrified into silence by the unexpected turn of events. Then a woman shrieked, and the spell was broken. Almost every one of the occupants of the deck-chairs stood up and rushed to the side, shouting as if noise would help the two men struggling for their lives.

The lascars too seemed incapable of action. They flocked to the side of the ship, and gazed seemingly without emotion into the deep-blue water.

At the shout of "Man overboard!" raised by Anstey, the officer of the watch, Captain Bullock unceremoniously dashed between the groups of bewildered passengers and gained the bridge. Even in his haste his brain was solving a ready problem. Who was to go away in the lifeboat? The Acting Chief was struggling for dear life in the "ditch". He could swim well, as the Old Man knew, but after his strenuous wrestling bout had he sufficient strength to keep afloat until picked up? Anstey, as officer on duty, could not leave the bridge. There was one executive officer short of the ship's complement, and as far as Captain Bullock was aware, none of the engineers off duty was capable of managing a boat, while a bungler at the tiller meant not only delay but probably failure.

Fortunately the *secuni* in the wheelhouse had acted promptly, putting the helm over to port in order to swing the ship's stern clear of the men in the ditch, and thus avoid the danger of their being cut to pieces by the propeller. They were now a good four hundred yards astern, while between them and the ship was a line of lifebuoys thrown with fine indiscrimination by the passengers. The nearest lifebuoy to the two exhausted men was at least a hundred yards away.

During the interrupted revels the *West Barbican* had reduced speed, and already Anstey had rung down for "Stop".

"Let go the lifeboat—away lifeboat's crew," bawled the Old Man, as he moved the telegraph indicator to full speed astern; then, leaning over the bridge rails, he hailed a grotesquely garbed figure standing motionless and alert on the temporary dais:

"Mr. Mostyn: take charge of the lifeboat."

With a feeling of elation Peter rushed to carry out the order. This time there was no question of it. The Old Man had spoken. It was a tribute to the Wireless Officer's capabilities in a province that was not strictly his own.

Urged by the shrill cries of the serang and tindal of the watch the lascars had now formed up on the boat-deck. Some had then their places in the out-swung boat, while others stood by the falls ready to lower away.

Although the engines had been going full speed astern the *West Barbican* was still forging ahead when Peter jumped into the stern-sheets of the lifeboat. She was still carrying way when the falls were disengaged and the boat pushed off from the ship's side.

"Soft job this," soliloquized Mostyn. "The sea's calm, the water's warm, and old Preston and the other fellow have got hold of the lifebuoy. Tumbling into the ditch under these conditions is a picnic—Hello, though—is it?"

*****

To say the least of it, Preston was both surprised and indignant when he found himself hurtling through space in the vice-like grip of his antagonist. It was poor consolation to know that there was someone else in the same predicament. What was particularly galling was the fact that he, a veteran officer of the Mercantile Marine, should be such an ass as to skylark and then fall overboard in so doing.

These thoughts flashed through his mind during the time he dropped through thirty odd feet of space between the deck of the ship and the surface of the water. Then the terrific impact with the Atlantic Ocean abruptly ended his reveries of self-reproach.

To a certain extent it was fortunate that the two men remained interlocked during their fall. Hunched up after the manner of a diver doing a "honey-pot" from a spring-board they got off comparatively lightly, although the impact was fairly severe, and had the effect of depriving them of most of the scanty breath left after their strenuous encounter.

"The blighter will grip like grim death," thought Preston, as he sank fathoms down; "I'll have a deuce of a job to shake him off."

But the sudden immersion had the unexpected result that the men mutually released their grip. Perhaps it was that both were good swimmers and realized that the quickest way to refill their lungs with air was to strike out for the surface.

They emerged almost simultaneously, gasping and spluttering.

"Not that way!" exclaimed Preston breathlessly, as his companion in misfortune began striking out for the ship's side. "Mind the prop."

The other realized the danger of being caught by the swiftly moving blades of the screw, but even then it was only the prompt action of the *secuni* at the wheel that saved him from being drawn into the vortex.

"Nothing to worry about," spluttered Preston, as the two bobbed like corks in the quartering wave. "We'll be picked up all right. My aunt! Look at them! Well, they might have chucked them on our heads."

He referred to the injudicious volley of lifebuoys. Although the ship was carrying way the passengers were still engaged in dumping the Company's property into the sea.

His companion laughed. Regaining his breath he was also regaining his boisterous spirits, although he had to admit that the struggle, followed by a thirty-odd foot fall had severely taxed his splendid brawn and muscle.

"You don't look in your element, Preston," he remarked, "even though you are Father Neptune."

"Was," corrected the absentee Acting Chief Officer, proceeding to relieve himself of the encumbrance of his scanty garb of trailing seaweed and oyster-shells. "Come on; we may as well strike out for the nearest of that line of lifebuoys. Breast stroke. There's no great hurry, and it's less tiring."

Although the passenger had gone overboard wearing boxing-gloves, that had remained on his hands despite his wrestling bout, one had disappeared during his submergence. Preston remarked on it.

"Yes," rejoined the other. "Might just as well hang on to this one, although one's not much use. Cost me a couple of Bradbury's just before we left England. I say, do you mind telling me this: I declare I've crossed the Line without being initiated. Is that so?"

"It is," replied Preston feelingly. "If you'd gone through the thing tamely we wouldn't have been in the ditch. Why did you ask me?"

By this time both men had swum to the nearest of the far-flung line of lifebuoys, and, glad of the support, were hanging on lightly at opposite sides of the buoyant "Kisbie".

"'Cause I want corroboration. Last night Murgatroyd bet me a tenner I wouldn't escape it. Have I won?"

"You have."

"Right-o, Preston!" was the delighted response. "I'll stand you a dinner in the swankiest hotel in Adderley Street as soon as we arrive at Cape Town. That's a deal. Hello! They're lowering a boat. What are you looking at?"

The Acting Chief Officer had seen the boat being swung out, and was calculating how long it would take to reach the spot where the lifebuoy was—calculating whether the boat's crew would find only an unoccupied lifebuoy floating in a patch of blood-stained sea—for less than fifty yards away was the black, triangular dorsal fin of an enormous shark.

"Nothing much," replied Preston, as calmly as he could, although the strained expression of his eyes was sufficient to attract his companion's curiosity. "Kick as hard as you jolly well can. Make a splash."

"Shark, eh?" exclaimed the co-partner of the life-buoy. "Right-o! I'm having my money's worth this trip anyway."

"Splash, man, splash!" was Preston's only rejoinder.

*****

"By Jove, I guess I look a sketch," thought Mostyn, as he steered the lifeboat towards the two men clinging to the buoy.

He certainly did. Called away hurriedly, he still wore part of his disguise as Amphitrite, Neptune's Queen. He had cast off his flowing locks of tow, but his well-powdered face and a vivid patch of rouge on either cheek looked absolutely grotesque. His costume of muslin (lent by one of the lady passengers) had suffered horribly during his attempt to squeeze through the hatch, while the trimmings of seashells and seaweed added to the weird appearance of the young Wireless Officer. To facilitate his movements Peter had "gathered in the slack" of his trailing garments, since without assistance he could not tackle the numerous safety-pins that his dresser had used in order to make sure that "nothing would come adrift and carry away".

"Hello, though—is it!" he reiterated, shading his eyes with his left hand.

Right in the glare reflected in the water his keen eyes had spotted a tell-tale swirl. Then above the surface appeared an object that settled his doubts. It was the dorsal fin of a shark.

One of the lascars, looking over his shoulder, saw the danger too. He raised a shrill cry that had the effect of startling his fellow-oarsmen and putting them off their stroke.

"*Chup rao!*" (Shut up), shouted Peter sternly. "Pull like blue blazes."

"Blue blazes" was evidently a stranger to the lascars' vocabulary, but they understood the word "pull" and guessed the significance of the rest.

Redoubling their efforts, they made the heavy boat travel rapidly through the calm water; but Peter realized that if the shark attacked with any promptitude the rescuers would be too late. He saw that Preston and his companion in distress were doing the best thing they could in the circumstances—making a violent splash. Whether the shark would be scared away was a matter for speculation.

Evidently the tiger of the deep was hungry. He was not devoid of pluck, for he had begun to swim round and round the two men, the while drawing nearer to the buoy. At any moment he might make a dart straight for his victims.

Peter knew this. He had seen a shark seize a South Sea Islander from a crowd of natives splashing and shouting in the surf. He had seen another monster seize and devour a dog within ten yards of a boat putting off to the animal's rescue.

There was no rifle in the lifeboat. In the Royal Navy they do things differently from the Mercantile Marine. Peter had an automatic. It was one of the things he took good care to provide himself with after his experiences in S.S. *Donibristle*; but the weapon was locked up in his cabin, and in the present circumstances it was like the Dutchman's anchor.

The boat was now a hundred yards from the life-buoy—the shark ten. The brute was still circling, sometimes diving, sometimes showing its head; but up to the present it had shown no sign of preparing to seize its prey by turning on its back.

A sudden inspiration flashed across Mostyn's mind. In the stern-sheets of the lifeboat was a box containing amongst other things a Verey's pistol. It was a weapon not of offence but for humane purposes. It was fired by means of a cartridge, but, instead of a bullet, it sent up a vivid coloured light to a height of about two hundred feet.

Peter stooped and opened the lid of the box. Thank Heaven! The pistol and cartridges were there. Deftly he opened the breech and thrust home the cardboard cylinder containing the detonator and explosive light; then, standing on the stern bench and steadying the tiller with one foot, he levelled the short-barrelled weapon.

For some seconds he waited. The shark in its orbit was immediately between the lifebuoy and the boat. Preston and his companion were in as much danger from the pistol as they were from the shark.

The huge fish dived and soon reappeared, this time well to the left of the buoy. It had partly turned on its back, and its wide-open jaws, triple lines of pointed teeth, and greenish-white belly were clearly visible, for by this time the whaler was less than twenty-five yards away.

It was now or never. The shark was preparing to make a dash for its victims under the bows of the boat.

Deliberately Peter pressed the trigger. He had to guess for elevation, knowing nothing of the trajectory of the missile. His aim was good. The rocket must have disappeared down the capacious maw of the shark, for there was no sign of the fiercely burning rocket sizzling on the surface. The satisfactory part of the business was that the shark disappeared and was seen no more.

Quickly the two men were hauled into the boat, both bordering on a state of collapse. Then, ordering the lascars to give way, Mostyn steered for the *West Barbican*, picking up the jettisoned lifebuoys on the way. He was one who always finished a job properly.

CHAPTER XV
Unpopularity

A few days later Mostyn was having an easy time. He was on watch, but with little to do. A notice-board on the promenade-deck furnished the reason for his enforced inactivity:

"S.S. *West Barbican*. To-day, in radio communication with *nil*. To-morrow, radio communication expected with *nil*."

The notice was painted with the exception of the two *nils*, which were written in chalk. Placed for the convenience of passengers wishing to send off private wireless messages, it duly recorded what ships and shore stations were within radio range. In her present position in the South Atlantic she was too far away to dispatch or pick up messages from Cape Town, the radius of her wireless being limited to 240 miles by day and almost thrice that distance by night.

Peter had overhauled the set, and was taking the opportunity of writing home. With his white patrol-coat unbuttoned and his *solar topee* perched on the back of his head, he was making the best of things in spite of the terrific heat and the attentions of numerous cockroaches.

There were thousands of these insects all over the ship, ranging in size from an eighth of an inch to nearly three inches in length. Whilst the *West Barbican* was in home waters their presence was invisible. They kept to the dark and inaccessible parts of the ship; but directly the weather grew warmer, as the ship neared the Tropics, they emerged fearlessly from their lairs and swarmed everywhere. By this time the passengers had grown more or less accustomed to them, but the early stages of the invasion of the living pests of the ship had caused great consternation and indignation, especially on the part of the ladies on board.

In times of boredom, when the passengers were "fed up" with deck-quoits and sweepstakes on the "day's run", the cockroaches would be pressed into service to provide entertainment. A dozen or more would be captured and placed on the deck, each having its own particular "fancier" in a miniature race, and it was surprising to see with what zest the passengers entered into the sport.

Presently Peter heard a light footfall on the deck, followed by a distinct knock upon the wide-open door of the cabin.

Rising, Peter found that Olive Baird was standing outside the brass-rimmed coaming.

"Good morning, Mr. Mostyn," she said. "Will you mind telling me if a message can be sent to Cape Town? And how much per word, please?"

"Sorry, Miss Baird," he replied, "we aren't in touch with any shore station. We may possibly get the Cape Town one to-morrow night."

At the back of his mind Peter found himself wondering why Miss Baird hadn't gone to the trouble of reading the announcement on the notice-board. He was rather glad she hadn't—perhaps she had purposely ignored it. It gave him an opportunity of entering into conversation with the girl.

Already Anstey had found out quite a lot about Olive Baird. How, he refused to divulge, but it was pretty certain that the girl had let out little or nothing.

Olive Baird was motherless. Her father had married again to a woman only five years older than his daughter, and, instinctively scenting domestic trouble in the near future, Olive had determined to earn her own living—a task that she had already found to be far more difficult than the cultured girl had imagined.

Almost at the end of her resources—for she knew that she would receive neither sympathy nor help from her estranged parent—Olive remembered a distant relation, a girl but a few years older than herself, who had married an official holding an appointment in the Kenya Colony.

To her Olive wrote, asking if there might be any post open to her in the district. Three months elapsed before the reply came—that there was a warm welcome awaiting her. Enclosed was a banker's draft, enough, and only enough, to pay for her passage out and to provide a necessary and simple outfit.

Before the *West Barbican* was many days out Mrs. Shallop, in one of her few amiable moods, had asked the friendless and reserved girl if she would, for a small remuneration, give her a couple of hours a day for the purpose of reading to her.

"My eyes aren't what they were," explained Mrs. Shallop. "And it's deadly dull on this ship when I can't even read."

So Olive thankfully accepted the post, because it helped her to pay her way; and, even when Mrs. Shallop had her almost at her beck and call, the girl did her best to keep on good terms with her.

It was not long before Olive found out the true nature of her supposed benefactress. Mrs. Shallop was vain, boastful, and with no regard for veracity. She was one of those persons who, having told the same fairy tale over and over again, firmly believe that the lie is the truth. On the other hand, her memory was defective, with the result that very frequently her story had a totally different setting when told a second or third time. In addition, she was bitingly sarcastic, and was never known to say a good word about anyone but herself.

So Olive had rather a rotten time.

The girl was, however, absolutely loyal to her employer. In the course of conversation with other passengers she was careful not to say a word that might be detrimental to Mrs. Shallop. Evidently that lady thought she might, for Argus-like she kept a strict watch upon her.

The Shallops had taken "Round Trip" tickets. These were issued by the Blue Crescent Line, and guaranteed a voyage of not less than three months. If by any chance, as was frequently the case, the voyage was prolonged, the holder of the ticket scored, for he or she was maintained at the Company's expense until the ship returned home or the passengers transferred to another vessel of the Company's bound for England.

Olive Baird's employers had made a heap of money during the Great War, and were now doing their best to spend it. Nevertheless, they wanted value for their outlay, and the round trip in the *West Barbican* pointed that way. Mr. Shallop was not keen on the voyage. It was his wife who insisted upon it, mainly because it was "the thing" to travel, and it would be an easy matter on their return to give out that they had gone on a palatial P. & O. mail-boat. It sounded grander than the Blue Crescent Line.

By this time the heat was beginning to tell upon the portly Mrs. Shallop. There were actually long intervals in which her strident voice failed to lacerate the ears of her fellow-passengers.

This was one of them. Wanting to do "the thing" and send a wireless message to her sister in Cape Town, Mrs. Shallop was too fatigued to mount the bridge-ladder; her husband had sheepishly slunk away to the smoking-room, and only Olive was available to undertake the commission.

"I'm sorry to have interrupted you," remarked Olive.

"Not at all; don't mention it," protested Peter; then, in an outburst of candour, he added: "You haven't seen our wireless-room."

"I should love to," rejoined Olive, who had the modern girl's leanings towards anything of a scientific nature. "I always wanted to see what it was like and how it worked, but I didn't like to ask you."

Without more ado Mostyn proceeded to explain the mysteries of that steel-walled house, unconsciously launching out into an intricate technical lecture on wave-lengths, atmospherics, induced current, valve and spark-gaps, until Olive was quite bewildered.

"There's nothing doing," he remarked, after the girl had placed the telephone ear-pieces to her shapely ears. "We're too far away from land. But I'll disconnect the aerial and let you see a ripping spark."

"Another time, Mr. Mostyn," demurred Olive. "Mrs. Shallop will wonder what I've been doing."

Calling silent maledictions upon the head of the tartar, Peter escorted the girl to the head of the bridge-ladder, extorting a promise that she would pay another visit to the wireless-cabin when the ship got within radiographic range.

"Or earlier if you like," he added.

He watched her disappear from sight and slowly made his way back to the cabin. Somehow the home-letter proceeded slowly and disjointedly. He was thinking of the jolly little girl who took such an interest in wireless.

Poor Peter! If he had only known how he had tired her almost to the verge of boredom.

Ten minutes after Miss Baird's departure Mostyn "got busy". Away to the starboard a vessel was calling CQ. The note was very faint and considerably hampered by atmospherics.

He was still endeavouring to tune in to the correct wave-length when he was interrupted by a vigorous punch between the shoulder-blades. Over his shoulder he saw that the interrupter was Mrs. Shallop.

Peter was rather more than annoyed by the interruption. He was angry. There was no denying that he possessed a temper, but he had usually the happy knack of keeping his feelings well under control. In the present circumstances he felt inclined to expostulate vehemently.

For one thing, he had a rooted dislike for the woman. For another, she had no right to be on the bridge, unless for the purpose of sending off a message or by the skipper's permission. Neither reason held just then. The wireless-cabin was closed for private transmission; she had not obtained the Old Man's sanction to be on the bridge.

The fact that Miss Baird had been on that spot only a few minutes previously hardly entered into Mostyn's calculations. Unconsciously he had allowed himself to be influenced by personal considerations, and he had forgotten that what was sauce for the goose was sauce for the gander.

With a deprecatory gesture of his left hand Mostyn attempted to convey the impression that he was busy. His attention had to be concentrated on the CQ message if he were to understand its import. It was difficult enough, without his being hampered by external interruptions.

One would have thought a hint sufficient. Not a bit of it! Mrs. Shallop was one of those hidebound, overbearing individuals who expected immediate and subservient attention.

"Why did you refuse to send off my message?" she demanded, in her loud, grating voice. "You put Miss Baird off with a trivial excuse, but that won't work with me, young man. Isn't my money as good as anyone else's? Don't you know that I'm the daughter of a naval — —"

Mostyn removed the telephones and stood up. There was an ominous glint in his eyes. His forbearance was nearing the breaking-point.

"I can only refer you to the notice-board on the promenade-deck," he said. "That and the intimation that passengers are forbidden on the bridge except with the Captain's permission. If you have any cause for complaint, please report to Captain Bullock. I must ask you to leave the wireless-cabin at once."

Mrs. Shallop recoiled as if she had received a blow on the face. She had expected no opposition. The quiet, decisive, and deliberate tones of the young Wireless Officer had completely taken the wind out of her sails.

Without a word she turned and made straight for the Old Man's cabin, bursting in like a tornado.

Captain Bullock was being shaved by his servant. The sudden and unexpected entrance of the tartar caused Wilkins's attention to wander, with the result that a crimson streak discoloured the lather on the skipper's chin.

Captain Bullock had, according to his usual custom, decided to remove his beard when approaching the Cape, and the operation was well advanced when Mrs. Shallop intruded at a very inopportune moment.

She failed to recognize the skipper shorn of his beard and with his face plastered with soap.

"Where's the Old Man?" she demanded heatedly.

What was the exact nature of Captain Bullock's reply Mostyn was unable to hear. With his mouth full of soap and his chin bleeding profusely the Old Man's articulation was a trifle confused; but he certainly did let himself go, with the result that the interrupter, in spite of her oft-reiterated claim to be a lady, was unceremoniously requested to remove herself to a region considerably warmer than the skipper's cabin, the temperature of which was registering 130° in the shade.

Chuckling to himself, Peter saw the discomfited Mrs. Shallop descend the bridge-ladder with more haste than dignity; then he tried, but in vain, to pick up the interrupted CQ signal.

"Captain Sahib him want you, sahib," announced Mahmed.

Mostyn promptly obeyed the summons. He too was rather surprised at the alteration effected by the removal of the skipper's beard, the newly shaven portion contrasting forcibly with the brick-red tan of the rest of his face.

"Tell me," began the Captain, "what was that old barge doing in the wireless-cabin?"

Peter explained.

The Old Man nodded eagerly.

"You did the right thing, my boy," he remarked "I've had enough— more than enough—of that impossible woman. I told her that in future she is not to come on the bridge on any pretext whatsoever. If she wants to send a message, let her; but she must do so in writing and submit it to me before it is passed. That'll clip her wings. All right, Mr. Mostyn, carry on."

Peter carried on until relieved by Watcher Plover. The latter was improving considerably, although he could never become an operator. He lacked the education and intelligence necessary for the work, but by this time he was able to discriminate between various signals and to know the Morse call for the ship. Consequently Peter's watch below was not subject to numerous and unnecessary interruptions.

"Hello, Sparks!" exclaimed Preston, as Mostyn blew into the smoking-room. "So you've been up against it this time. Tell us all about it."

There were about half a dozen passengers, the Acting Chief Officer, and two of the engineers off duty passing a pleasant hour. All seemed eager to know full particulars of the encounter.

"She's an unmitigated nuisance," declared an artist, proceeding to Natal in order to paint some frescoes for one of the important buildings. "We'll all be reduced to nervous wrecks before we see the last of her. Can't we choke her off?"

"For Heaven's sake don't, old chap," protested Comyn, his cabin-mate, a tall, lean-faced, literary man. "I bear the brunt of it. Every morning I get a dose of it until I know every shred of her personal history in spite of the fact that the details vary as consistently as does the ship's position. It is priceless. I revel in it. Wouldn't miss it for worlds; I encourage her, in fact."

"'Tany rate," interposed Alderton grimly, "she called you a lanky reptile."

"Perhaps," rejoined the unruffled author. "If it comes to that, she said you were a little worm. There's no end of fun making out that you believe all Mrs. Shallop tells you. It's a little gold mine."

"For you, perhaps," added Preston. "However, I guess the Old Man has upset her apple-cart. We won't hear her bell-like notes again in a hurry."

But he was mistaken. Into the smoke-laden atmosphere wafted the strident voice of the lady under discussion. She was venting her wrath upon Olive Baird.

CHAPTER XVI
Hot Work in No. 1 Hold

The S.S. *West Barbican* was within a couple of days of Cape Town. The weather, although still warm, had lost much of the sweltering heat, thanks to the influence of the Trades.

The ship was rolling badly. For the last ten days she had been on her best behaviour in that respect; but now she was making up for lost time. There was a high sea running, and the ship's alley-ways to the saloon were ankle-deep in water.

With the glass falling rapidly the seas increased in violence. It was evident that the *West Barbican* would receive a heavy dusting within the next few hours.

"Hanged if I like the look of things, Preston," admitted Captain Bullock, sniffing the approaching storm from afar. "We're in for something."

"We are, sir," agreed the Acting Chief. "And I'm not altogether satisfied with that steelwork. Bad enough cargo at any time, but I've an idea something's working adrift in No. 1 hold. I'll get Anstey to have a look at it."

The Old Man concurred.

"Tell the serang to warn the lascars," he added. "We don't want broken limbs and all that sort of thing."

At an order a party of lascars assembled for the purpose of securing any of the cargo that might have broken adrift. Presently Anstey, wearing sea-boots, made his way along the lurching deck. He was not at all keen on this particular job. Hounding about in the semi-darkness of the hold and in momentary danger of being crushed by a mass of shifting metal was not a pleasing outlook. But it was duty, and Anstey was not a shirker.

The lascars cast off a portion of the tarpaulin and removed the aftermost of the metal hatches, disclosing the rusty coaming and the upper portion of a vertical ladder of iron—or, to be more precise, a ladder that was nominally vertical. In present conditions it was swaying with the ship, and describing an erratic curve with a maximum heel of twenty degrees.

Steadying himself by the coaming, Anstey felt with his left foot for the topmost rung. Then, gripping the sides of the ladder, he began the descent.

Very little daylight found its way into the narrow space afforded by the displaced hatch. In fact Anstey soon found himself in gloom approaching total darkness. The air too, after being confined for weeks, was dank and distinctly unwholesome. There was an acute smell from the fumes given off from the red oxide with which the steelwork had been coated.

With his rubber-soled boots slithering on the slippery rungs as the vessel rolled, and gripping strongly with both hands, the Third Officer descended until at length his feet came in contact with the metal floor of the hold. The din was terrific. Without, the seas were hammering on the comparatively thin hull-plating. Bilge-water was foaming and hissing in the cellular bottom, while the vibration of the engines—the noise intensified in the confined space—added to the turmoil.

To these noises Anstey paid scant heed. He was listening intently to a metallic sound, which told him that Preston's precautions had not been taken in vain. Somewhere in the for'ard part of the hold there was a regular metallic thud. It came from a mass of metal that had worked loose from the securing chains.

Anstey's first intention was to order a couple of lascars below.

"May as well do the jolly old job myself," he soliloquized, on second thoughts.

Fumbling in his pocket he produced his electric torch. For some minutes he was dazzled by the blinding glare. Then, as his eyes grew accustomed to the light, he could form a good idea of the difficulties of his surroundings.

He was standing in a narrow fore-and-aft passage. The walls consisted of red-painted girders piled up to a height of ten feet on either side of him. Although secured by chains and upright steel bars they presented a formidable appearance, as alternately each wall towered obliquely over his head, the whole mass straining and groaning at its lashings like a Titan striving to burst his bonds.

Staggering along the narrow passage, for the erratic movement of the hold was totally different from the heave and pitch to which Anstey was accustomed on deck, the Third Officer made his way cautiously forward, critically examining the metal gripes that secured the awkward cargo.

Suddenly he stopped. A cold perspiration stood out on his forehead. Danger, imminent danger, stared him in the face. Danger not only to himself but to the ship and her passengers and crew.

Three feet above his head a huge girder was chattering and quivering. The chain that secured it to its fellows had at one time been set up by a massive bottle screw. Possibly the thread was an easy one, but, in any case, the constant working of the ship had caused the bottle screw to "run back". It was now holding by a couple of threads at the most, and momentarily the securing chain might fly asunder.

Anstey realized what that meant. The fifty-ton girder would crush and pulp him to a jelly. Not only that; it would to a certainty start the bottom plates of the hull and shatter the bulkheads of No. 1 hold as well. That meant that the *West Barbican* would plunge like a stone to the bed of the Atlantic.

Thrusting the barrel of his torch under the strap of his peaked cap, Anstey replaced the headgear, jamming it on so that the peak was over his right ear. That gave him a direct light to work with.

Then, pulling out the marline-spike of his knife, and holding it between his teeth, Anstey began to scale the precarious wall of steel until he could tackle the almost disjointed bottle screw.

It seemed an eternity climbing that five or six feet. To his agitated mind it seemed as if the girders were already slipping bodily upon him. As his toes sought an insecure hold he could feel the steelwork trembling. With each lurch of the vessel to starboard the bottle screw strained, until the young officer felt certain that the last two threads had stripped and the last restraining bonds had been loosed.

At last he found himself in a position to tackle his task. With one foot resting on a girder on one side of the passage, and the other on the opposite side, and steadying himself as best he could with his left hand, Anstey inserted the point of the marline-spike in the slot of the bottle screw.

Then he began to turn the locking device, slowly and firmly.

HE BEGAN TO TURN THE LOCKING
DEVICE, SLOWLY AND FIRMLY

At first he was seized with the terrifying idea that the threads were not gripping. With the torch in his cap throwing its rays erratically with every movement of his head, Anstey felt convinced that his efforts were in vain.

He went on turning and turning, barking his knuckles as the tapering spike slipped again and again. Then, with a grunt of satisfaction, he saw that the ends of the threaded bolts had reappeared.

Even as he looked, the torch slipped from his cap and clattered to the metal floor. The hold was plunged into darkness.

His first impulse was to make for the open air. In the darkness the difficulties of working in the place were redoubled. It required a determined effort to force himself to his incompleted task.

Solely by sense of touch he carried on, until he had the joy of feeling the reunited ends of the threaded bars. That part of the business was finished until next time, he decided.

Regaining the floor, he felt his way between the piled-up girders until his hand came in contact with the ladder. Twenty-five feet above his head he could see a rectangular patch of light, one edge broken by the heads and shoulders of half a dozen lascars.

Up the ladder Anstey swarmed, drinking in copious draughts of the pure, salt-laden air.

But his task was incomplete. He must make sure that everything in No. 1 hold was secure.

"Thatcher, old son," he exclaimed, as he encountered one of the junior engineers. "Lend me your torch, there's a good sort. I've scuppered mine."

Thatcher fumbled in the pocket of his dungarees.

"Here you are, you careless blighter," he replied. "Skylarking, I suppose? Well, take care of my gadget, anyway."

Again Anstey descended the hold and completed his survey. The clang of shifting steel had ceased.

When, after an hour's absence, he regained the bridge, Preston was not to be seen, but the skipper spotted the dishevelled youth and sung out to him.

"Well?" queried the Old Man.

"All correct, sir," reported Anstey. "The——"

"Good," rejoined the Captain, without waiting for the Third's explanation. "Carry on."

Anstey turned away to "carry on". It was his watch below. The job in No. 1 hold was merely an extra. He was still feeling the effects of his desperate efforts in the confined space, and the idea of turning in before he had had a "breather" did not appeal to him.

On the lee side of the bridge he encountered Mostyn.

"Hello, old thing," was Peter's greeting. "What have you been up to? You look a bit green about the gills."

"Nothing much," replied Anstey. "Just been giving an eye to your father's ironmongery. Yes, it's all right. Got a cigarette? My case is down below. Thanks awfully."

CHAPTER XVII
The Decoy Wireless

The *West Barbican's* stay at Cape Town was of short duration. She landed about a score of her passengers and a small quantity of cargo, coaled, and proceeded, giving Peter little opportunity of a closer acquaintance with the oldest city of South Africa.

He was fairly busy during the run round to Durban, since the ship was within wireless range both of Cape Town and the seaport of Natal. Consequently he spent most of his waking hours in the wireless-cabin, rather than have to be continually called by Partridge and Plover.

The *West Barbican*, having spent a night at anchor under the Bluffs at Durban, proceeded alongside the quay to disembark the bulk of her passengers and a considerable amount of cargo.

It was here that "Mr. Porter" severed his personal acquaintance with the ship, although his interest in the *West Barbican* did not in the least degree wane. On the contrary it was rapidly increasing.

With a Kaffir porter carrying his portmanteau and suit-case von Schoeffer passed along the gangway and gained terra firma. He had found no suitable place in which he could secrete his explosives, nor had he an opportunity for so doing; so the only course that remained open, short of dumping the stuff into the sea, was to take it ashore with him.

He anticipated no difficulty in passing the Customs. None of the officials would detect in the harmless-looking slab that resembled sheet-glue one of the strongest explosives possible to obtain. They were "traveller's samples" and as such were allowed duty free.

So within ten minutes of leaving the *West Barbican* Ludwig Schoeffer was bowling along in a rickshaw, drawn by a huge, muscular Zulu "boy", en route for a small hotel that overlooked the harbour.

On the following day Schoeffer's explosive, with the detonator timed for its maximum limit, was stored in No. 3 hold of the S.S. *West Barbican*, as one of the twenty odd cases of hardware consigned by the well-known firm of Van der Veld to Senhor Perez Bombardo of Beira.

Simply but effectively disguised, Schoeffer saw the crate whipped on board and lowered into the hold. So far so good. It looked as if he were certain of success. He chuckled as he conjured up a mental picture of the head director of the Pfieldorf Company handing over a substantial cheque.

During the rest of the *West Barbican's* stay at Durban, Ludwig Schoeffer lay low. For the present he had done all that was necessary. His deep-laid scheme was progressing favourably.

His idea was to signal the ship by means of wireless and, by spurious authority, order her to Rangoon. It was not unusual for ships of the Blue Crescent Line to receive unexpected orders when on the high seas, since they held roving commissions once they were round the Cape and had landed their mails.

And, since it would take longer than the seven days to make Rangoon, the *West Barbican* would end her career mysteriously in mid-Indian Ocean.

At ten one morning the *West Barbican* stood out to sea bound for Beira and Pangawani, at which latter place she was to land the consignment of steelwork for the Kilba Protectorate.

At four the same afternoon Schoeffer walked into the offices of the wireless company at Durban.

"I want this message sent to the *West Barbican*," he announced, handing in a form written in code—the private code of the Blue Crescent Line.

The clerk accepted the form without demur. He had no idea of its meaning, nor had he any way of finding out. Not that he wanted to. Messages in code were the rule rather than the exception.

The message as received and ultimately sent off by the shore operator was as follows:

"SW. TLB. FEW. CNI. TLXQ. VP AELB TNI PU. AEMQ".

Ludwig Schoeffer paid the eighteen shillings demanded and obtained a receipt. Then, having got an assurance that the message would be dispatched within an hour, he wished the clerk good afternoon and walked briskly to the waiting rickshaw.

The bogus message read, when decoded:

"I have received telegraphic instructions from your owners for you to proceed straight to Rangoon, where you will unload steelwork, proceeding thence to Port Sudan".

CHAPTER XVIII
The Difference of a Dot

"Hello, Sparks; you look a bit off colour?"

This was Dr. Selwyn's greeting as Mostyn, having handed over the watch to Plover, walked into the doctor's cabin.

"I feel it, Doc," replied Peter. "Touch of the old complaint—malaria."

Selwyn had detected the symptoms the moment the Wireless Officer showed his face inside the door. Peter was trembling violently. He was feeling horribly cold, and his head was aching badly.

"Taken any quinine?" asked the medical man.

"Yes," was the reply. "My ears are buzzing already."

"Then turn in," ordered Selwyn. "I'll make you up a draught. Keep as warm as you jolly well can. This will make you perspire freely before midnight, and you'll be fit by this time to-morrow."

Peter waited while the doctor made up the medicine, and then staggered to his cabin, where Mahmed, greatly concerned, helped his master into bed and piled blankets and a bridge-coat upon his shivering body.

It was now one bell in the first dog watch.

At two bells Peter was still awake and trembling with cold spasms when Watcher Plover hurriedly entered the cabin.

Plover had no idea that Mostyn was down with malaria, and it was not unusual for him to find Peter lying on his bunk when off duty.

"Call for the ship, sir," he reported. "No bloomin' error this time. SVP as sure's my name's Plover."

Mostyn kicked off the blankets and rolled out of the bunk. He staggered as he stood up, and would have been glad of Plover's assistance. But the Watcher, having delivered his message, had gone back to his post.

With a terrific buzzing in his ears Peter almost dragged himself along the alleyway and up the bridge-ladder. Many a time he had regretted the absence of a second wireless officer. Now, above everything, he wanted an efficient substitute; but, of course, none was available.

Entering the wireless-cabin, he picked up the telephones and gave the acknowledgment. Then, a pencil in his trembling hand, he waited for the text of the message to come through:

"SW. TLB. FEW. CNI. TLXQ. VP AELD TNI PU. AEMQ".

Yes, Peter had that all right, but, ever on the cautious side, he asked for the message to be repeated.

"Here you are," he said, handing the duplicate message to his assistant. "Nip off with that to Captain Bullock."

"Don't you look rummy, sir?" remarked Plover, noting for the first time Mostyn's drawn features.

"Am a bit," admitted Peter. "I'll be all right by the morning. Skip along."

Watcher Plover "skipped along" at his usual stolid pace to the Old Man's cabin, while Peter, almost incapable of controlling his trembling limbs, somehow contrived to regain his bunk.

"Signal just come through, sir," reported Plover, as he handed the pencilled form to the skipper.

"All right," replied the Old Man brusquely. "Hand me that book; the second on the left. That'll do, carry on."

It did not take Captain Bullock long to decode the message, but a frown of perplexity spread over his forehead as he read the momentous words.

Then he rang the bell and ordered Plover to return.

"Who received this?" he asked.

"Mr. Mostyn, sir; he had the signal repeated."

"All right. You may go."

The assurance that the Wireless Officer had personally taken down the code message removed all doubts from Captain Bullock's mind.

"Mr. Preston," he sang out.

"Ay, ay, sir,"

"Fresh orders," announced the Old Man. "Here you are: 'I have received telegraphic instructions from your owners for you to proceed straight to Bulonga, where you will unload steelwork, proceeding thence to Port Sudan'. Bring me the chart of the Mozambique coast, Preston, and let's see where we are—and the sailing directions while you are about it."

The Acting Chief hastened to fetch the required articles.

"Bulonga—that's in Mozambique," commented the Old Man. "What the blazes the Kilba Protectorate people want to have the steelwork dumped there for goodness only knows. However, it's my place to carry out instructions, Mr. Preston."

"Ay, ay, sir," concurred the Acting Chief without enthusiasm. He had no love for the Portuguese East African ports. A long spell there meant mosquitoes; mosquitoes meant malaria and other evils in its train. And there was simply nothing to see or do in these ports. Preston had "had some" before to-day.

"They give no reason for the alteration, sir?" he inquired. "I suppose by any chance we haven't got the signal incorrectly?"

"No reason, Mr. Preston," replied Captain Bullock. "And here is the signal in duplicate. Mostyn took that precaution, so I can stake my boots on its accuracy."

The two officers spent some time in poring over the chart and reading up the description of Bulonga harbour and its approaches, as set down in the Admiralty sailing directions for the east coast of Africa.

"It'll be a tight squeeze for our draught," commented the skipper. "It'll mean a Portugee pilot, worse luck. I know those gentry of old. I hope there's a British agent there to take over the Brocklington Company's consignment."

Had Captain Bullock known that Peter was down with a severe bout of malaria he would not have wagered his footgear so readily, for Mostyn had made a mistake in taking in the signal. More, he had duplicated the mistake when he received the repetition at his own request.

With his head buzzing like a high-pressure boiler Peter had read D ($-$..) for B ($-$...), his temporarily disordered sense of hearing failing to detect the slight but important difference.

Consequently, instead of the *West Barbican* shaping a course for Rangoon, which in the code signal appeared as AELB, she was making for the comparatively unimportant harbour of Bulonga (AELD).

The while Ludwig Schoeffer's seven-day watch was silently ticking out the seconds, minutes, and hours in the *West Barbican's* baggage hold. The German agent was sublimely ignorant of the change in the ship's plans. He was still at Durban, awaiting the news that the *West Barbican* was overdue and believed missing. He would have been considerably surprised had he known that there was every likelihood of the ship sinking in Bulonga Harbour, where at low tide she would have barely enough water to lie alongside the quays.

If he had only known the vital difference that the omission of a "dot" in the spurious signal was to cause!

CHAPTER XIX
Peter's Progress

Peter Mostyn's attack lasted a full twenty-four hours, but at seven the next evening he felt well enough to go down to dinner in the saloon.

That function had become a mere shadow of its former self. On the run to Cape Town the chairs round the long tables were generally filled, once the passengers had grown accustomed to life afloat and had regained their temporarily lost appetites.

Now, the saloon looked almost deserted. Captain Bullock was in his customary place at the head of the table, most of the officers not on duty were present—a mere handful all told. Of the passengers only eight remained. Of these, five were to be landed at Beira and taken on to their destinations by a "Bullard" boat. The remaining three were Mr. and Mrs. Shallop and Olive Baird.

Since Mrs. Shallop's encounter with the skipper she had fought shy of the saloon when the Old Man was present, and was in the habit of having her evening meal in the seclusion of her cabin. Although this arrangement was contrary to the Company's rules and regulations Captain Bullock winked at it; the rest of the saloon congratulated themselves, and even Shallop, away from the disturbing influence of his wife's presence, seemed a different man. In fact, on several occasions his dry and somewhat humorous remarks set everyone laughing.

The temporary retirement of Mrs. Shallop had given Olive much more leisure. At first the selfish woman had tried her level best to compel the girl to share her self-imposed seclusion, but Olive had firmly declined to submit. She had already endured considerable discomfort on her employer's behalf, and had borne the almost continuous "nagging" without a murmur; but now the breaking-strain had been exceeded, and the bullying woman had to admit defeat.

Consequently Peter saw Olive a good deal. They were firm pals. There was nothing sloppishly sentimental about the girl. She was merely a jolly little person emerging from the temporary cloud of reserve caused by the depressing influence of the naval captain's daughter.

She had been fully initiated into the mysteries of the wireless-room; she had taken equal interest in the complicated machinery of the engine-room; and, since leaving Cape Town, Captain Bullock had given her permission to go on the bridge whenever she wished. She had coaxed Anstey into showing her how to "shoot the sun" and to use the *Nautical Almanac* in order to work out the ship's position. Even the *secuni* in the wheelhouse so far forgot his duty as to allow the Missie Sahib to take the wheel.

But undoubtedly her interest was keenest in sailing. Both Preston and Anstey had promised to give her a run in one of the *West Barbican's* sailing-boats while the ship was at Durban. This promise they severally performed, but to a certain extent the beat to windward and the run home on the spacious but shallow water of the harbour was a disappointment to Olive — since neither man had offered to let her take the tiller.

Dinner over—Peter had very little appetite—Olive Baird went on deck, and somehow, whether by accident or design, Mostyn found her standing on the starboard side of the promenade-deck, gazing at the moon as it rose apparently out of the Indian Ocean.

"What a topping evening, Mr. Mostyn," observed the girl. "Just fine for a sail."

She gave a glance at one of the quarter-boats, an eighteen-foot gig fitted with a centre-board.

"'Fraid it can't be done," remarked Peter, with a laugh. "Stopping vessels in mid-ocean for the purpose of giving lady passengers a spin in one of the boats isn't usual. Might work it when we arrive at Bulonga. You're fond of sailing, I notice."

"I love it," declared the girl enthusiastically. "Do you?"

"Yes, rather," agreed Peter; "so long as there's not too much of it."

"There never could be too much as far as I am concerned," protested Olive. "What do you mean by too much?"

"Well, for instance, a two-hundred mile run in a boat of about that size," replied the Wireless Officer, indicating the centre-board gig. "I tried that sort of thing once, but the boat never reached her destination."

"Tell me about it," commanded Miss Baird. "Were you single-handed?"

"No," replied Peter. "There were three fellows and a girl. We got wrecked."

For nearly three-quarters of an hour Olive listened intently to Mostyn's account of the escape from the pirate island in the North Pacific; the narrator with his natural modesty touching but lightly upon his share of the desperate enterprise.

"And where is the girl now?" inquired Olive.

"She married my chum Burgoyne," replied Peter. "I had a letter from him when we were at Cape Town. Burgoyne is a jolly lucky fellow."

"We had a sailing-boat of our own once," said Olive, her mind going back to those far-off days before she had a stepmother to make things unpleasant for her. "I used to sail quite a lot on the Tamar when we lived at Saltash."

"Bless my soul!" exclaimed Peter to himself. "I felt certain I'd seen her before, but I couldn't for the life of me say where."

For a few moments he remained silent, making a mental calculation.

"Was it in 1913?" he inquired. "Didn't you have a bright, varnished boat with a teak topstrake and a red standing lugsail? And you were about eight or nine then. You used to have your hair bobbed, and wore a white jersey and a scarlet stocking cap?"

"However did you know that?" asked Olive in astonishment.

"Because we had a yacht moored just above the red powder hulks. My father held an appointment at Keyham Dockyard, you see; and whenever he had a home billet he kept a yacht or boat of some sort. Sailing was his favourite pastime."

But Olive was paying scant heed to the description of Mostyn *père* as set forth by Mostyn *fils*. Her thoughts too were flying back to those halcyon days before the war.

"I believe I remember you," she said at length. "Weren't you on board a white yawl of about six tons, with a green boot-top and rather a high cabin top?"

"That was the *Spindrift*, my pater's yacht," declared Peter. "And ——"

"And you were about ten or eleven, with a freckly face," pursued Miss Baird calmly. "You were a horrid little wretch in those days, because I distinctly remember you laughing at me when the halliard jammed and I couldn't get the sail either up or down."

"Guilty, Miss Baird," said Peter. "I apologize. Give me a chance to make amends and I'll be all over it."

"I will," agreed the girl. "You may take me for a sail in Bulonga Harbour; but you mustn't be selfish, like Mr. Preston and Mr. Anstey. You will let me take the tiller, won't you?"

Peter gave the required promise. He felt highly pleased with himself. Anstey was evidently in disfavour because he had underrated Olive's capabilities as a helmswoman. In addition, the Third Officer would be fairly busy while the *West Barbican* was in harbour, as the steelwork had to be taken out of the hold. Reminiscences of youth spent in the West Country, too, were mutual and sympathetic bonds between the Wireless Officer and the girl. No wonder he was feeling highly elated.

"What sort of a place is Bulonga?" asked Olive.

"Haven't the faintest idea," replied Peter. "Never heard of the show until a day or two ago. Don't expect a second Durban, Miss Baird. If you do you'll be disappointed. I shouldn't be at all surprised if it's a pestilential mud-hole. By Jove, it's close on eight bells, and it's my watch."

Half an hour later Mostyn "took in" a message from Durban addressed to Miss Baird. It contained the brief announcement that Mr. and Mrs. Gregory—Olive's relations to whom she was on her way—were returning to England in three days' time, and that Olive's passage-money home was lying at the Company's offices at Durban.

CHAPTER XX
An Eventful River Trip

"What a one-eyed crib!" exclaimed Anstey, as the *West Barbican* slowly approached the low-lying coast in the neighbourhood of Bulonga.

Mostyn nodded in concurrence.

The outlook was dreary in the extreme. All there was to be seen was a squalid collection of galvanized-iron huts rising above a low, sandy spit; a few gaunt palms; a line of surf—not milk-white, but coffee-coloured—and a background of sun-dried hills.

The whole coast seemed to have been scorched up by the sun. Brown and drab colours predominated. The foliage was of a sombre drab-green narrowly approaching a dull copper colour. Even the sea in the vicinity of the harbour had lost its usual clearness and appeared to be charged with a muddy sediment.

"Any sign of the pilot, Mr. Anstey?" inquired Captain Bullock.

The "S International", the signal for a pilot, had been flying from the topmast-head for the last hour, as the *West Barbican* cautiously closed with the inhospitable-looking coast, but there were no signs of activity ashore.

In ordinary circumstances it was customary for the ship to wireless her agents, asking them to make arrangements for a pilot; but, since there were no agents at Bulonga, nor even a wireless station, that procedure was put out of court. There remained only the old-time flag signal to summon a pilot from shore.

"No sign yet, sir," replied the officer of the watch. He had been scanning the shore through a telescope until his eyes smarted. The glare form those "tin" huts seemed to be reflected through the lenses of the telescope to his optic nerve. He was literally seeing red.

"All asleep, I suppose," commented the Old Man. "It beats me why we've been ordered to this rotten hole. Try 'em with the siren, Mr. Anstey."

The echoes of the powerful whistle had hardly died away when a hoist of bunting rose slowly in the humid air. Until a faint zephyr caught the flags it was impossible for the *West Barbican* to understand the import of the signal.

"FWE," sang out Anstey. "That reports that there's not enough water on the bar, sir."

"Not enough fiddlesticks!" snapped the Old Man. "It's within half an hour of high water. We'll lose the flood if they don't get busy. Besides, how the blazes do they know our draught? For two pins I'd take her in myself."

No doubt the skipper, with the aid of chart, compass, and lead-line, could have navigated the ship across the bar with complete success. He had worked his way into uncharted harbours before to-day. But should the vessel ground he would be in a very difficult position with the Board of Trade. Even if he were successful in getting the ship safely alongside the quay there might be trouble with the Portuguese officials for not complying with the port regulations.

"That chap who wrote something about those serving who only stand and wait didn't know much about the tides," fumed the Old Man. "Here's the blessed tide serving, but it won't stand and it won't wait, and time's precious."

Nevertheless the skipper had to wait, impatiently and irritably, until such times as the easy-going officials sent out a pilot.

It was more than an hour later before a white motor-boat with an awning fore and aft was seen approaching the ship.

As the boat drew nearer its ugliness became apparent. The paint was dirty, and in places rubbed away to the bare planking. The awning had seen better days, and had been roughly patched in a dozen places. A couple of coir fenders trailed drunkenly over the side, while the painter was dragging through the water. The motor was wheezing like a worn-out animal and emitting smoke from numerous leaky joints, while the clutch, slipping badly, was rasping like a rusty file.

A Zanzibari native was "tending" the engine, and a half-caste Portuguese was at the wheel. In the stern-sheets was a short and very stout man puffing at an enormous cigar. He wore a dirty white uniform with a lavish display of tarnished gilt braid, while set at an angle on his bushy hair was a peaked cap with the Mozambique arms.

"Goo' mornin', Senhor Capitano!" he exclaimed, when the boat ranged awkwardly alongside. "Me pilot. Get you in in shake o' brace—no—brace o' shake."

Still puffing his cigar the Portuguese pilot came over the side and waddled on to the bridge.

"Vat you draw?" he inquired.

The Old Man gave him the ship's draught.

"Ver' mooch," rejoined the pilot, shrugging his shoulders. "Tide go. Why you no call me before?"

But get her in he did, although the propeller was throwing up muddy sand and the keel plates were slithering over the bottom.

Half an hour later the *West Barbican* was berthed alongside the quay—a dilapidated structure partly stone and partly timber, with rusty bollards that, judging by their appearance, had not made the acquaintance of mooring-ropes for months. Clearly the maritime activities of Bulonga were largely dormant.

Presently—there was no hurry, everything at Bulonga being done on the "do it to-morrow" principle—the Customs officers came on board.

They were bilious-looking rascals, whose broad hints for "palm-oil" were as plain as the fellaheen demanding baksheesh. To them the task of searching for dutiable goods was of secondary importance.

From one of them, who spoke English passably, Captain Bullock elicited the information that there was no British agent in the place; neither was there telegraphic, telephonic, nor railway communication with anywhere. Once a week a small steamer brought up outside the bar for the purpose of collecting and delivering mails and parcels. When the weather was rough, or the bar impassable, the inhabitants of Bulonga had to wait another week, perhaps two, for news of the outside world.

"We'll have to hand over the steelwork to some one, Preston," observed the Old Man. "We can't dump it on the quay and leave it to rot. Nip ashore and see if there's a fairly reliable storekeeper who will freeze on to the stuff till it's wanted. We'll need a covered store at least a hundred and twenty feet in length."

The Acting Chief returned on board with the information that there was a suitable place, and only one. The owner, a timber exporter and importer, had gone home, and no one knew when he was likely to return. He lived at a place called Duelha, about seven or eight miles up the river that empties itself into the shallow Bulonga Harbour, and he was in the habit of journeying to and fro by means of a motor-boat.

"We'll have to rout him out," decided Captain Bullock. "I'll send my motor-boat. Meanwhile we'll engage natives and start getting the stuff out of the hold. The question is: who am I going to send away with the boat? You'll be on duty on deck, Preston, and Anstey will be tallying in the hold. I've got it. I'll get young Mostyn to go."

He went to the end of the bridge and looked down. On the promenade-deck were Peter and Olive watching the dreary harbour.

Miss Baird had taken her great disappointment remarkably well. On the principle that there is no time like the present, she refused to dwell upon the prospects of returning home. She would have to, she supposed, in due course; meanwhile she was on board the *West Barbican* without any immediate chance of returning even as far as Durban. And the longer the voyage the better, she decided.

"This doesn't look promising for our sail, Miss Baird," said Peter. "The tide's ebbing like a millrace. Look at those trunks of trees coming down. They'd give a small boat a nasty biff if they fouled her."

"And no wind," added the girl. "Mr. Preston was telling me that in the harbours on this coast it blows from the land from sunset till about ten o'clock, and from the sea from a little after sunrise till ten in the morning. Between times it's usually a flat calm."

The harbour viewed from within looked far more uninviting than it did from the offing. The ebb was in full swing—a turgid, evil-smelling rush of coffee-coloured water. Already the mud-banks fringing the mangrove-covered islands were uncovering and throwing out a noxious mist under the powerful rays of the tropical sun, which was now almost immediately overhead.

Mostyn found himself comparing Bulonga Harbour most unfavourably with the lovely lagoons and coral reefs of the Pacific islands.

"It may be better later on in the afternoon," he remarked. "Say an hour before high water. If——"

He broke off abruptly, for Captain Bullock was descending the bridge-ladder.

"Hello, young lady!" exclaimed the skipper. "What do you say to a run in my launch? I'm sending her up-stream in a few minutes. You'll be snug enough under the double canopy over the stern-sheets."

"It ought to be rather exciting, Captain Bullock," replied Olive, glancing at the surging ebb. "It would be very nice to see what it's like."

"Right-o!" rejoined the skipper. "Mr. Mostyn, will you take charge of the boat? You seem the best man for the job, considering that it's your father's steelwork we are dealing with. Take this letter to a Senhor José Aguilla, who hangs out at a place called Duelha. I'll show you it on a chart. Get him to come down as soon as possible. If he's like the rest of these gentry that will be *mañana*. In any case, bring back a written reply to this letter."

"Very good, sir."

"Carry on, then. Pass the word for the serang to have the motor-boat hoisted out and the awnings and side-curtains spread. Miss Baird, can you be ready in a quarter of an hour?"

Mostyn hurried away to carry out his instructions.

"Good sort, the Old Man," he soliloquized. "And at one time I thought I'd hate him like poison. It just shows a fellow that it's not wise to judge by first impressions."

Promptly the serang and half a dozen lascars came upon the scene and began to cast off the lashings that secured the motor-boat to No. 2 hatch. The little craft was Captain Bullock's private property. She was about twenty-five feet in length, carvel-built of teak, and had a 12-horse-power paraffin engine installed under the fore-deck. 'Midships was a well, fitted with a wheel and motor controls, while the spacious cockpit aft was provided with a folding hood, as well as double awnings spread between tall brass stanchions.

In less than ten minutes the boat had been swung out by means of a derrick, and was straining at her painter alongside the accommodation ladder.

With Senhor Aguilla's letter in the breast-pocket of his drill tunic and his automatic in his hip-pocket, Mostyn waited at the head of the ladder until Olive appeared, wearing a light, linen skirt and coat and a topee with a gold-edged pugaree.

It was "stand easy". Notwithstanding the tremendous heat the officers were spending their leisure in a manner followed by Britons all the world over. They were playing cricket, with the netted promenade-deck as the field, and stumps precariously supported by a small wooden base. Yet the thrill of deck-cricket paled into insignificance when Olive Baird appeared. One and all the players flocked to the side to watch her departure in the Old Man's motor-boat.

From the top of the accommodation ladder Peter signed to the native engineer, and by the time Olive stepped agilely into the stern-sheets, without taking advantage of Mostyn's proffered hand, the motor was purring gently.

"Let go aft—let go for'ard!" ordered Peter. "Touch ahead."

By a gentle movement of the wheel Mostyn got the boat clear of the ship's side without the risk of hitting the propeller. He knew from experience that the effect of helm is to swing a boat's stern round and not her bows. Then, with a sign to the native engineer to "let her all out", Peter steadied the boat on her course.

The Old Man's private launch was no sluggard. She could do a good nine knots, but her progress against the formidable ebb seemed tediously slow. She was slipping through the coffee-coloured water quickly enough, as her bow-wave and clear wake denoted; but she seemed to be crawling past the low river banks at less than a slow walking pace.

Peter did not mind. He had no idea of wasting time in the execution of his orders, but, on the other hand, the relatively slow progress did not worry him. He was perfectly happy. Olive, too, was obviously enjoying the run. The breeze set up by the motion of the boat through the still air was delightfully cooling after the enervating atmosphere on board the *West Barbican* alongside the wharf.

"Like to take her?" asked Peter, when a bend of the river hid them from the ship.

"Rather," replied Miss Baird promptly, and, nimbly negotiating the bulkhead between the stern-sheets and the steering-well, she mounted the low, grating-fitted platform and grasped the wheel.

Mostyn, who had relinquished the helm, stood just behind and a little to the side, so that he could command a view ahead. Occasionally he had to consult the chart in order to avoid the numerous sand-banks.

"Look out for those floating logs, Miss Baird," he cautioned, as three or four huge tree trunks, green with trailing weed, rolled lazily over and over in their aimless passage to the open sea.

Olive avoided them easily. Peter's confidence in the helmswoman increased by leaps and bounds. There was no hesitation on her part, no bungling as the swift, frail craft passed between two of the logs with less than six feet to spare on either side.

"Give that log a wide berth, Miss Baird," observed her companion, after a number of obstructions had been avoided. "Unless I'm much mistaken we'll find that log has a motor of sorts. Yes, by Jove! it has!"

The "log" was an enormous hippopotamus, floating motionless on the water, with only its snout and a small portion of its back showing above the surface.

At this point the river had contracted considerably, the actual waterway being less than twenty yards from bank to bank, although at half tide these banks were submerged and the width of the stream increased to nearly a quarter of a mile.

Olive meant to give the brute as wide a berth as possible, while, on the other hand, the hippo resolved on close quarters with the motor-boat.

Instead of diving to the muddy bottom of the river the hippopotamus began to swim rapidly towards the launch, opening its huge jaws with evident relish at the prospect of biting out a few square feet of gunwale and topside as an entree.

Mostyn and the native coxswain, who had hitherto been "standing easy", were keenly on the alert. The latter, seizing an oar, made ready to deal a blow upon the brute's head, although the hippo would have paid no more attention to the blow than he would to being tickled with a straw.

Olive showed no sign of nervousness. In fact, she acted so coolly and with such excellent judgment that Peter made no attempt to grasp the wheel.

Seeing the animal approach, the girl edged the boat well over to the port side of the narrow channel. In spite of the speed of the launch it was apparent that the hippo would cut it off if the same direction were maintained.

Not until the boat's stem was within twenty yards of the brute did Olive alter helm. Then, with a quick, even movement, she put the helm hard-a-port.

Before the unwieldy animal could turn, the launch had literally scraped the hippo's submerged hindquarters. Then, swinging the boat back on her former course, the girl glanced at her companion.

"Near thing, that," she remarked. "I wonder that would have happened if we'd hit it?"

"We would have come off worst," replied Peter, who, now the danger was over, was beginning to realize what the consequences might have been.

"Perhaps you wouldn't mind taking on," said Olive a little later.

Mostyn took the helm. Although the girl had given no reason for wanting to relinquish the wheel, he felt pretty certain that the incident had shaken her up a bit.

"You're all right?" he asked.

"Quite," was the reply.

Presently the river widened considerably. The launch was now within half a mile of her destination, but, according to the chart, there was a submerged bank on the starboard hand, and fairly deep water close to the right bank.

Without warning the impetus of the launch was arrested. Peter was flung against the wheel; Olive, losing her balance, cannoned into him, and was saved from a violent concussion against the coaming by the fact that the native coxswain had got there first, and had been winded by his impact with the woodwork. The engineer, who had crawled under the fore-deck to replenish the contents of a grease-cup, was flung along the narrow floor by the motor and finished up by butting the petrol tank.

"Aground!" exclaimed Mostyn, stating what was an obvious and accomplished fact.

The engine was racing furiously. Jerking the reverse lever into the astern position Peter hoped that the action of the powerful propeller would release the launch from her predicament. It was in vain. The motor was racing as fast as ever, but there was no flow of water past the boat's side to indicate that the propeller was going astern.

"Blades stripped, by Jove!" ejaculated Mostyn.

He switched off the ignition, and, in the relative quietude that succeeded the machine-gun-like explosions of the exhaust, took stock of the situation.

"Quite all right, thank you," replied the girl, in answer to Peter's question. The reply set Mostyn wondering whether in any circumstances Olive would say otherwise.

By this time the native coxswain was sitting up. Although he was not taking nourishment he was gently caressing the bruised part of his anatomy, but otherwise betraying no interest in things.

Then the engineer appeared, backing out of the motor-room, and mopping the blood on his forehead with a silk scarf. Gaining the steering-well he drew himself up and salaamed.

"Why sahib stop engine?" he inquired.

"'Cause the propeller blades are gone," replied Mostyn. "Savvy? Blades—screw—no can do. Like this."

He tried to convey the magnitude of the disaster by means of dumb show. The native failed to understand. Being aground mattered little to him; being slung about like a pea in a box he took more or less as a matter of course. The thing—the thing that counted—was the fact that the sahib had taken unto himself the duty of Abdullah Bux, engineer of the Sahib Captain's launch, and had stopped the motor. Abdullah Bux felt that on that account he had a grievance.

The launch was lying well down by the head in about a couple of inches of water. Her stem had struck a waterlogged tree trunk almost buried in the soft mud. The impact had lifted her bows well clear of the water, the greater portion of the keel passing over the obstruction until, the bows dropping and plunging into the mud, the boat came to a standstill. Then it was that the swiftly moving propeller had fouled the log, with the result that the three blades were shorn off close to the boss.

"Tide still ebbing," remarked Peter. "We're properly on it, Miss Baird."

"Yes, unfortunately," was the rejoinder. "There's no way of getting her off till the tide makes?"

"Might try kedging her off," suggested Mostyn.

"A kedge wouldn't hold in this slime," declared the practical Miss Baird, "even if you were able to lay it out. But you can't. The mud's too soft."

Peter sounded with an oar. The blade sank almost without resistance to a depth of three feet in the noxious slime.

A tedious wait followed. There was no denying the fact that it was tedious. Peter and the girl sat under the after canopy, but a *tête-à-tête* under these conditions was very different from one on the promenade-deck of the *West Barbican* on a tranquil, starlit night. It was hot—insufferably so. Not only did the sun pour fiercely down upon the double awning. The mud, now "dry", was radiating heat—a clammy, evil-smelling heat, as the rotting vegetation left high and dry by the receding tide lay sweltering in the sunshine. The heavy, motionless air, for there was not the faintest suspicion of a breeze, reeked as only the air of an African swamp can— an overpowering, nauseating stench. Thrown in as a makeweight came the reek of hot oil from the badly overheated engine.

"Tide's turning," said Peter, breaking the long silence.

There was no lull in the change from ebb to flood. At one moment the brownish waters were foaming seawards; at the next a miniature "bore" was breaking over the fringe of the mud-flats, bringing with it a collection of flotsam in the form of branches and trunks of trees.

"'Fraid I'm giving you a rotten time," continued Peter apologetically. "Sailing with Preston and Anstey in Durban must have been a joy compared with this—and you told me you didn't like it a bit. You must think I'm a rotten pilot."

"Nearly everyone gets aground some time or other," replied Olive. "The awkward part is that this isn't exactly like the mud-banks of the Tamar. And it's unfortunate about the propeller. What do you propose to do when we float?"

"Row up to Duelha. It's less than half a mile. If we can't get a spare propeller we might ask Senhor Aguilla to tow us back in his motor-boat."

The flood-tide made with great rapidity. In less than half an hour the launch was afloat. The two lascars manned the oars, and the boat, borne rapidly by the tide, quickly covered the remainder of the way to Duelha.

The Portuguese agent was overwhelmingly polite. He insisted on entertaining Olive and Peter to coffee, and promised to tow the disabled launch back to the ship, at the same time regretting that there were no facilities at Duelha for repairs.

"Eet is no trouvel, senhor," declared the Portuguese. "I myself vill speak to el capitano Bullock concerning de stores from de sheep. Eet is pleasair to do business vid de Englees all de time."

It was sunset before Olive and Peter returned to the S.S. *West Barbican*.

CHAPTER XXI
The End of S.S. *"West Barbican"*

Throughout the day the scantily clothed Bantu workmen had been busily engaged in unloading the steelwork. The natives, unlike their Portuguese masters, had to keep hard at it, with the result that by the time "knock-off" was announced and the Bantus, resuming their calico skirt-like garments, had trooped ashore, the S.S. *West Barbican* drew five feet less for'ard than when she crossed the bar. Captain Bullock's interview with Senhor José Aguilla was of a mutually satisfactory nature. The latter undertook to store and look after the consignment of the Kilba Protectorate until such time as it was claimed by the authorities. The terms were so many thousand milreis per month, a sum that on paper looked truly formidable, but actually was equal to about seven pounds of English money.

The Old Man was pleased to get the steelwork off his hands so reasonably. Senhor Aguilla was pleased because he had the steelwork on his hands. That was the difference.

The Portuguese knew that the longer the consignment remained unclaimed the longer he would continue to draw a fairly substantial sum for wharfage and storage; and, although he promised to forward a letter to the Kilba Protectorate agent at Pangawani by the next weekly steamer, he meant to take steps to prevent, for as long as he possibly could, the information concerning the steelwork reaching the proper quarter.

Having, as he thought, satisfactorily settled with Senhor Aguilla Captain Bullock sent for his Wireless Officer.

"That means a ticking off, I expect," thought Peter, when Mahmed delivered the message. "The Old Man's rattled about his motor-launch."

Mostyn was only partly right in his surmise. Captain Bullock was annoyed, which was natural enough. No boat-owner likes to have his craft damaged, especially when he is not on board. He has a sort of feeling that the accident, whatever it might be, would not have occurred had he been present. It was an awkward mishap. Until the *West Barbican* returned to

Durban, or some other large port, it would be hopeless to expect to obtain a new propeller.

But the skipper, in spite of his bluntness, was a just man. He dealt with cases impartially, and no one having been censured by him had good reason to doubt his judgment.

Peter went to the skipper's cabin and reported the circumstances of the accident. The Old Man listened attentively until the Wireless Officer had finished his narrative; then he pointed to a chart of Bulonga Harbour that was lying on the desk.

"Show me where the stranding occurred, Mr. Mostyn. What, there? On the port-hand side of the channel?"

"Yes, sir."

Captain Bullock had no cause to doubt Peter's word, but he made up his mind to question the two lascars who were in the boat, and also to see if Miss Baird could throw any light upon the matter.

"H'm. I suppose the river has changed its bed," he remarked. "African rivers have a nasty habit of doing that. It was unfortunate that you struck a snag; otherwise it wouldn't have mattered very much. All right, carry on."

Abdullah Bux and his compatriot could give no definite information. Miss Baird, for the present, was not available. The strident tones of Mrs. Shallop indicated pretty clearly that the lady was bullying the girl for her prolonged and involuntary absence.

At sunrise next morning the *West Barbican*, drawing considerably less water than she had done eighteen hours previously, recrossed the bar. The Portuguese pilot was dropped, and a course steered to pass through the broad Mozambique Channel. Without exception all on board were glad to get away from the malodorous harbour of Bulonga.

On the afternoon of the seventh day after leaving Durban the weather "came on dirty". A heavy wind from the east'ard raised a nasty sea, which would have been angry but for the torrential downpour of rain that had the effect of beating down the crested waves.

As darkness set in the sky was almost one continuous blaze of vivid sheet lightning. The rain was still heavy but the wind piped down, blowing softly from the nor'-east.

"We haven't seen the last of this yet," declared Preston. "The glass is a bit jumpy. It'll blow like billy-ho before morning. How about your aerial, Sparks? Aren't you going to disconnect it?"

The two officers, clad in oilskins and precious little else, were keeping the first watch. There was nothing doing in the wireless-cabin. Atmospherics were present, but, apart from these disturbances, no sound had been audible in the telephones during the best part of Peter's watch. Insufferably hot, he had put on an oilskin and had gone out for a breather.

"No need," he replied. "At least not until we get forked lightning."

"I'm not sorry we've got shot of that steelwork," remarked the Acting Chief after a pause. "It's awkward stuff to carry. But the trouble of it is that removing it has altered our deviation. The compass cannot possibly be the same with that enormous amount of metal taken out of the ship. I suggested to the Old Man that we ought to have swung the old hooker before we left Bulonga and adjusted compasses. But he was in a hurry to get under way, and, apart from that, the harbour was so shallow that we couldn't get a clear swing. She's not far out on this bearing. I took a sight at the Southern Cross for that. Talking of compasses: did you hear that yarn about the Flinder's bar?"

"About the candidate for Mate's certificate who told the examiner that: 'There ain't no pub o' that name in Gravesend'?" asked Peter.

"No, but that's not so dusty," replied Preston. "My yarn concerns an old skipper in the Penguin Line. He was— —"

But Mostyn was not to hear the anecdote.

A violent concussion, as if the ship had struck a rock, almost threw the two men off their feet. A muffled report followed.

"Mined, by Jove!" exclaimed Preston, in the brief lull that succeeded the detonation.

Then pandemonium was let loose. The lascars, yelling and shouting, poured on deck, followed by a mob of native firemen. Capable enough in ordinary circumstances, the Indians lacked the stolidity and grim courage of British crews when disaster, sudden and unexpected, stared them in the face.

Captain Bullock was quickly on the bridge. He could do little or nothing to allay the panic, for the native petty officers were as frantic as the rest. To add to the difficulties of the situation, every light on board went out. Vast clouds of smoke and steam were issuing through the engine-room fiddleys.

The propeller was slowing down. The engineer on watch had, on his own initiative, cut off steam and opened the high-pressure gauges.

The Old Man shouted through the speaking-tube to the engine-room. There was no response.

Just then, in the glare of the lightning, he caught sight of Anstey, who, awakened by the explosion, had hurried to the bridge in his pyjamas and uniform cap.

"Nip below, Mr. Anstey, and see the extent of the damage," he ordered.

Anstey turned to obey. At the head of the bridge-ladder he encountered Crawford, the engineer of the watch.

"Nice sort of night to be in the ditch, laddie," exclaimed Crawford, as he elbowed his way past the Third Officer. "How far is to land, anyway?"

Crawford was on his way to report to the bridge. He had been flung violently on the bed-plates when the explosion occurred. Upon regaining his feet he found the engine-room in darkness save for the feeble glimmer of an oil lamp. Water was pouring in like a sluice through a rent in the after bulkhead that separated the engine-room from No. 3 hold. The firemen, panic-stricken, were bolting on deck. Neither by words nor action could Crawford stem the human tide of affrighted Asiatics.

Quietly he made his way to the platform and awaited orders from the bridge. The telegraph remained silent, the indicator on the dial still pointing to "Full Ahead".

By this time the water in the stokeholds was damping the fires, and Crawford deemed it prudent to shut off steam and open the escape valves in order to avert an explosion of the boilers.

Knee deep in the oily water that slushed to and fro as the ship rolled, the engineer of the watch groped his way through clouds of steam until his self-appointed task was done. Then, after shouting in case anyone else had remained below, he effected his retreat and at once made for the bridge to report to the Old Man.

"She's going, Mr. Preston," declared Captain Bullock.

"She is, sir," agreed the Acting Chief. Experience had taught him the now unmistakable symptoms of a foundering ship.

"Call away the boats," continued the Old Man "If you've trouble with that mob use your revolver, Preston. Don't hesitate. Remember we've women on board. Use your discretion as to what boat you stow 'em in."

The Acting Chief hurried off, pausing outside the wireless-room to give Mostyn the last known position of the ship, which information was a necessary adjunct to the SOS call.

Peter had not been idle. The moment the seriousness of the situation became apparent he was back at his post in the wireless-cabin.

The shutting off of steam had automatically stopped the dynamo. In any case, the explosion had severed the "leads". The main set was out of action. Mostyn had to fall back upon the emergency gear.

For quite ten minutes he contrived to call up, but no reassuring reply came through in reply to the urgent appeal for aid. There were ships within range of the emergency set, that Peter knew. He had spoken them earlier in the evening.

"Either atmospherics or else they've another Partridge and Plover on board," he thought grimly. "Wonder where my birds are?"

The two Watchers ought to have been on the bridge by this time. In case of distress it was their duty to "fall in" outside the wireless-cabin and await instructions. Neither had done so.

The floor of the cabin had quite an acute list by this time. It was only by propping his legs against the lee bulkhead that Mostyn could keep seated. He realized perfectly well that the ship was sinking rapidly, but it is part of an unwritten code of honour that a wireless officer "stands by" until he is ordered away by his skipper or swept from his post by the sea itself.

Even as he waited, still sending out the unacknowledged SOS, he thought of Olive Baird, wondering how she was faring in the horrors of the night. If he only knew—but perhaps for his peace of mind it was as well that he did not.

Above the turmoil without came the report of two pistol shots in quick succession. There was no mistaking the sharp cracks. They differed completely from the detonations of the distress rockets that at intervals were fired from the bridge, on the chance that a vessel in the vicinity might proceed to the aid of the foundering ship.

The pistol shots reminded Peter of something that he might otherwise have overlooked. Without removing the telephones from his ears he groped and found his automatic and a box of cartridges.

"No knowing when it might come in useful," he soliloquized, as he thrust the weapon into his hip pocket. "While I'm about it I might as well get dressed."

With considerable difficulty, owing to the now terrific list of the ship, he contrived to throw off his oilskin and don his white patrol suit over his pyjamas. Then, putting on his oilskin once more, he waited.

He had not much longer to wait.

"Any luck?" inquired the Old Man, who was gripping the doorway of the wireless-cabin with both hands in order to prevent himself slipping bodily to lee'ard.

"No, sir," replied Mostyn.

"Then chuck it," continued the skipper. "Look nippy. She's nearly gone. Where's your life-belt?"

A slight recovery on the part of the stricken *West Barbican* enabled Peter and the skipper to gain the weather bridge rail, the former securing a lifebelt from the chest by the side of the chartroom.

It was a weird and terrible sight that met Mostyn's eyes as he clung to the rail. The vivid flashes of lightning threw the scene into strong relief as the bluish glare illumined the night.

Not only was the ship listing to port. She was well down by the stern, her poop being practically submerged. From the lee side of the boat-deck a row of empty davits overhung the black water, the lower blocks of the disengaged falls flogging the ship's side like a series of blows with a sledge-hammer.

A cable's length away was one of the boats with only half a dozen people in her. Another more laden was a little distance away, the rowers laying on their oars. A third, deep in the water, was laboriously putting away from the ship. A fourth, waterlogged, with her bow and the top of the transom showing above the surface, was drifting at some distance astern of the ship, while a fifth was floating bottom upwards with five or six lascars struggling to clamber upon the upturned keel.

"We'll have to shift for ourselves, Mostyn," said the Old Man calmly. "The best of luck!"

The people in the sparely manned boat, noting the skipper and the Wireless Officer on the bridge, began to back towards the foundering ship.

"Avast there!" bawled Captain Bullock. "Stand off. Keep clear of the suction. She's going!"

With a shudder like an animal in mortal pain the staunch old ship made her final plunge. Amidst the rending of wood, as the enormous pressure of confined air burst the decks asunder, and the crash of the funnel as the guys carried away, she slid stern foremost beneath the waves.

Then a violent rush of water swept Peter off the shelving planking of the bridge. He was conscious of being flung heavily against some solid object, turned round and round like a slowly spinning top, and being dragged down, down, down.

Vainly he tried to keep his breath. The pressure on his lungs became intolerable. He was barely conscious of struggling madly in the crushing embrace of the black water.

Then everything became a blank.

CHAPTER XXII
A Night of Horror

Acting Chief Officer Dick Preston, on receiving the Old Man's order to get the boats away, lost no time in getting to the scene of operations. The frantic rush of the lascars to the boat-deck warned him of what to expect. He had seen the panic-stricken clamour of a crew of white-livered dagoes, had watched them tumble pell-mell into the sole remaining boat, and had witnessed the result—a swamped whaler and twenty men struggling for dear life, and struggling in vain in the icy cold water off the Newfoundland Banks. That was many years ago, but the lesson had not been lost on Dick Preston.

Hurriedly loading his revolver, the Acting Chief gained the boat-deck. Already the native crew had swung out one of the boats, and a fierce struggle was in progress between the lascars and the firemen as to who should go away in her.

There was no love lost between the two classes. They were of different races, the lascars hailing from Bombay while the firemen were recruited from the Coromandel coast; they were of different faith, the former being Mahommedans, the latter Buddhists. It needed little to cause a row. When it came to a struggle for life the natives were in a state bordering upon madness.

"Chup rao!" shouted Preston, levelling his revolver. "Belay there! Stand fast!"

For a moment the lascars and firemen hesitated. Then, as the ship shook and staggered as the bulkhead of No. 2 hold gave way, they surged in a living torrent into the out-swung boat, regardless of the revolver shots which the Acting Chief fired over their heads.

Preston made no further attempt to restore order on the boat-deck. If the men disobeyed orders he was no longer responsible for their safety.

He passed along until he came to a knot of comparatively amenable Madrasis, who had been gathered together by Anstey and two of the engineers.

"Right-o, old man!" exclaimed the Acting Chief to the Third Officer. "Lower away! You take command, and good luck to you."

Quickly, yet with good discipline, the boat was manned and lowered — Anstey, the two engineers, and Mr. Shallop in the stern-sheets.

"Keep in company, Mr. Anstey," shouted Preston, as the falls swung free.

"Ay, ay, sir," was the reply, followed by the order: "give way."

Anstey's boat was barely clear of the side when the first boat to be swung out was let go with a run. Greatly overcrowded, it struck the water with tremendous force. The impact broke her back, and in a moment she filled, leaving the frantic natives floundering in the water. Some were crushed as the sea flung the waterlogged craft against the ship's side. Others strove to clamber into the boat, only to destroy her slight buoyancy. In the mêlée knives were used with deadly effect, until only half a dozen men, who had swum clear of the boat, were left out of the thirty odd who had crowded into her.

It had been both Preston's and Anstey's plan to get the women away first; but each had quickly realized that this was out of the question. For one thing, neither Mrs. Shallop nor Olive was on the boat-deck. For another, it was useless to attempt to place them in the boats until the panic-stricken mob was effectively dealt with.

Two more boats, each under the charge of an engineer, and with three or four stewards, got away with difficulty. The crowd on the boat-deck had thinned considerably.

"Now, then, where are the women?" demanded Preston. He was not altogether certain whether they had already got away, for, save for the less frequent flashes of lightning, the scene was in total darkness.

"Here you are, Preston!" shouted a voice that the Acting Chief recognized as the Purser's.

A bluish glare, a prolonged flash, enabled Preston to see the missing passengers. The Purser was literally dragging Mrs. Shallop along the deck, while Olive was close behind.

For once Mrs. Shallop was silent. She was unconscious.

"I wondered why she wasn't complaining that she was not being treated as a lady," thought Preston grimly. "That accounts for it."

Together, the Acting Chief and the Purser unceremoniously bundled the insensible woman into the last boat but one on the port side. Those on the starboard were useless, for, owing to the excessive heel, they could not be lowered clear of the sloping side.

"Now, Miss Baird."

Guided by Preston the girl entered the boat, in which were three lascars—one of them Mahmed, Peter's boy.

"Where's Mostyn?" shouted the Acting Chief. "Partridge! Plover! Hurry up, now!"

He called in vain. The two Watchers had already got clear of the ship. Mostyn was still vainly endeavouring to get the SOS message through.

Meanwhile the Purser, the Chief Steward, and the remaining natives had lowered the last available boat. Preston was left alone on the boat-deck—a fact that was revealed to him when the next lightning-flash rent the sky.

"Where's the Captain?" he shouted, hailing the boats lying a short distance away. "Anyone seen Captain Bullock?"

By this time the water was washing over the well-deck. At any moment the *West Barbican* might turn turtle.

A voice from one of the boats replied:

"Here!"

"What's that?" bawled Preston.

"All right," answered the voice.

The Acting Chief was puzzled. It was not the Old Man's voice, but perhaps Captain Bullock had been injured. He had not seen the skipper since he left him on the bridge. Apparently the bridge was deserted. It looked untenable owing to the great list of the ship.

A muffled explosion, as yet another bulkhead gave way under the pressure of water, warned Preston that it was time for him to go. It was his duty to take charge of the boat in which were the two women passengers.

Leaping into the boat, Preston signed to Mahmed to help him with the after falls, at the same time shouting to the other two lascars to lower away handsomely.

Although there was no one on deck to man the falls, it was a fairly easy matter to lower away the comparatively light boat with only six persons on board, the distance from the davit-heads to the water being only about ten feet, so deep had the ship settled.

"Fend off!" ordered Preston, as he jerked the lever of the patent disengaging gear.

Even as he spoke the heavy metal block of the lower after falls swung violently outwards. In the darkness the Acting Chief did not see the impending danger.

The next instant the swaying lump of metal struck Preston full on the temple. Without a groan or a cry he pitched headlong upon the stern-sheet gratings.

It was Mahmed who discovered the apparently lifeless form of the Chief Officer. He communicated his discovery to his compatriots, and an excited conversation ensued. Meanwhile the boat was drifting aimlessly at less than ten yards from the West Barbican's port quarter. Until it occurred to the lascars—who were arguing on a question of precedence as to who should now give orders—that there was imminent danger of the boat being swamped by the suction of the foundering ship, they made no effort to man the oars.

When about a hundred yards from the ship the lascars ceased rowing and resumed their argument.

All this time Olive had done what lay in her power to render Mrs. Shallop's plight less painful. She was in utter ignorance of the accident that had befallen the luckless Acting Chief Officer, although she was rather puzzled at the lack of discipline displayed by the lascars, and the fact that the officer in charge of the boat made no attempt to check the dispute.

Another vivid sheet of lightning illumined the scene, but Olive was not looking into the boat. Her attention was attracted by the sight of two men standing on the listing bridge of the ill-fated West Barbican.

The glare was of sufficient duration to enable her to distinguish Captain Bullock and Mostyn. She saw the former raise his hand and beckon the boat to pull clear. He was shouting something, but in the turmoil the words were indistinguishable.

The long-drawn lightning flash ended, leaving the girl blinking in Stygian darkness.

"There's Captain Bullock and Mostyn still on board, Mr. Preston," she exclaimed, in anxious tones. "Can't we put back to fetch them?"

There was no reply.

In a louder tone Olive repeated the question of entreaty.

Still there was no answer.

The lascar bowman resumed his oar, pulling the boat's head round. Finding his companion idle he prodded him in the back with his foot, with the result that the man gave a few desultory strokes. In the utter darkness the lascars had lost all sense of direction, and, instead of pulling away from the ship, they were slightly closing with her.

Suddenly a hissing sound rent the air. It was the ship plunging beneath the waves. The boat, caught by the turmoil of the tempestuous seas, was thrown about like a cork. One of the men was hurled off the thwart by the loom of his oar striking him in the face. The oar was swept from his grasp and lost overboard.

To Olive, crouching on the bottom-boards, it seemed as if the boat were being lifted vertically. The movement reminded her of the sudden and unexpected starting of a lift. Then, heeling terribly, the boat dipped her gunwale under, and a cascade poured into her until Olive was sitting waist deep in water.

Her first act was to raise Mrs. Shallop's head. The shock of the water had caused that lady partly to recover consciousness. She was moaning and coughing.

The violent motion lasted for quite a minute, then the maelstrom subsided, and the partly waterlogged boat bobbed sluggishly on the waves. The lascars, now roused to activity, were baling furiously with their hands, since in the darkness it was impossible to find the baler which was supposed to be in the boat.

"Mr. Preston!" exclaimed Olive once more.

"Preston Sahib he dead man," was Mahmed's startling announcement, although the words were delivered with the imperturbability of the Asiatic.

The horror of the situation gripped the plucky girl. Throughout the period between the explosion and the foundering of the *West Barbican* she had been perfectly self-possessed, her chief solicitude being for her tyrannical employer. Now the full magnitude of the disaster became apparent. She and the unconscious Mrs. Shallop were alone in the boat with three apparently incapable lascars. Preston was, presumably, dead; Mostyn she had seen standing on the bridge just before the ship sank, keeping up the traditions of the Wireless Service to remain at his post as long as the ship was afloat and the transmitting apparatus was capable of being worked.

The other boats were neither to be seen nor heard. Whether they were still standing by or whether they were making for the nearest land the girl knew not.

She would have welcomed another lightning flash, out none came. The electrical storm had passed. Rain was now falling heavily, and the total absence of wind was ominous. It presaged a hard blow, possibly a storm, at no distant date.

Olive was thinking deeply. It was "up to her" to show the lascars that a British woman is not helpless in a tight corner.

"If only it were light," she thought.

Then she remembered that the boats usually carried an emergency equipment, an oil lamp amongst other things.

"Mahmed," she ordered, "get the boat's lamp from the stern-locker and light it."

She would have found it herself, but for the fact that Preston's body lay on the stern-gratings. She frankly admitted to herself that nothing could induce her to grope her way past that in the darkness.

The two lascars were still baling in the bows. They too were reluctant to go aft, where, by removing the stern-sheet gratings, they could deal more effectually with the water in the bilges.

Mahmed obeyed without protest. Olive could hear the search in progress; first the clatter of the detached locker-cover, as it slipped upon the stern-sheets, then the rasping of a metal-bound keg, and the metallic clank of the lamp.

"No can do, memsahib," reported Mahmed. "No light, no match."

"Look again," commanded the girl. Unless some unprincipled person had purloined them, there ought to be matches in a watertight box along with the rest of the gear in the after locker.

A further search proved futile. The boats and their gear had been inspected by the officer of the watch only that morning, and had been reported as being in good condition and fully equipped in every respect. Either Anstey, as inspecting officer, had shirked his whole duty or else, which to Olive seemed unlikely, the matches had been stolen in broad daylight.

"See if there are matches in Preston Sahib's pocket," said the girl.

But Mahmed drew the line at that. In his quaint English he explained, giving several reasons that seemed puerile.

"I suppose it's hardly fair to get him to do what I daren't do myself," thought the girl. Then, summoning up her resolution, she leant over the stroke-thwart, and shudderingly groped for the Acting Chief's pockets.

To her delight she found a box of Swedish matches in the breast pocket of Preston's drill patrol jacket. Before she could withdraw her hand the supposedly dead man moved slightly, but none the less perceptibly. That altered the situation. Olive was no longer dealing with a corpse, but with a living person. Instinctively she placed her hand over Preston's heart. It was beating very feebly.

"Here are matches, Mahmed!" she exclaimed. "Light the lamp quickly. Preston Sahib is not dead."

It seemed an interminable delay before Mahmed succeeded in getting the lamp lighted. The matches were damp, the wick wanted trimming, and the colza oil was a long time before it gave out a flame.

At length the lamp was lighted, and there was quite a steady light, and the transition from utter darkness imparted confidence.

Giving a hasty look at Mrs. Shallop, to see that she was still in the recovering stage, Olive turned to the more important work in hand.

Preston looked a ghastly sight. One side of his face had been badly injured, while the concussion had caused blood to ooze from his eyes, nose, and mouth.

Olive's first step was to wash the injured man's face and moisten his lips with water. She had the good sense to use salt water for the washing process, knowing that the contents of the water-beaker were likely to be more precious than gold before the adventure was over. Then, pillowing the patient's head on a sail and covering him with a piece of tarpaulin, she debated as to what was to be done next.

Clearly Preston's case required medical aid. Selwyn was in one of the boats, but whether they were in company or not Olive had no idea.

"Hold up the lamp, Mahmed," she ordered. "High up."

The boy obeyed, while Olive, shading her eyes from the heavy rain, peered around in case any of the other boats might be displaying a light. It was a doubtful point. Even if they had, the torrential downpour would tremendously curtail the range of visibility of the low-powered light.

In fact, held high above Mahmed's head, the rays simply illuminated a circular patch of rain-threshed water, a little more than a dozen yards in radius, Beyond was an impenetrable wall of darkness.

An involuntary cry came from Olive Baird's lips. She could hardly believe the evidence of her eyes, for floating inertly within an oar's length of the boat was a man—Peter Mostyn.

Whether he was alive or dead Olive knew not. His usually tanned features looked a ghastly greenish hue, his eyes were closed, and his head was hanging sideways. His arms were moving slightly, but the movement was purely automatic as the lifebelt-clad figure lifted to the gentle undulations of the sea.

Startled by Olive's cry, Mahmed looked in the direction to which the girl was pointing. His fright at seeing, as he thought, the dead body of his master, was almost disastrous in its result. The upheld lamp slipped from his nerveless fingers and fell clattering upon the gunwale. For an instant it seemed uncertain whether it would drop into the sea or not, but luckily a movement of the boat slid it inboard.

But the fall had extinguished the lamp. Mahmed was in too much of a blue funk to relight it. Olive settled the question by taking the box of matches from him and lighting it herself.

Neither of the two lascars for'ard would move a finger to row towards the Wireless Officer. Superstition akin to panic held them in its grip. They would not—they could not—use their oars. Every bit of courage seemed to have oozed out of them.

Seizing one of the spare oars lying across the thwarts, Olive, using the unwieldy ash paddle-wise, slowly brought the boat nearer and nearer the seemingly inanimate man. Had there been any wind the task would have been almost impossible, owing to the high freeboard of the lightly laden boat; but in the absence of even a faint breeze Olive was able to accomplish her aim.

With a sigh of relief she threw down the oar, and, leaning over the gunwale, grasped Peter by one arm.

CHAPTER XXIII
Peter takes Charge of Things

Exerting every ounce of strength, Olive tried and tried in vain to haul Mostyn into the boat. In normal conditions he was no light weight, and now, in his waterlogged clothing and wearing a cumbersome lifebelt, he was so heavy that the girl could do no more than lift his head and shoulders clear of the water.

She called to the lascars for assistance, but the only reply she received from the two men for'ard was: "No good; him dead man."

Mahmed, however, although he had no doubt that he was handling a corpse, came to her aid, although he worked with an averted face. Even with his assistance Olive had a hard task, but at length Peter was unceremoniously bundled over the gunwale, and placed in the stern-sheets close to the unconscious Preston.

Anxiously the girl gazed at his pallid face, hoping to detect some sign of life. Then she began the operations as laid down in the instructions for restoring the apparently drowned.

In her schooldays Olive had been taught this useful knowledge, but she had never before had an opportunity of putting the knowledge to the test. She felt none too sure of it. Once or twice she found herself wondering whether she was doing the wrong thing.

For a full half-hour she kept up the respiratory exercises, until, in the uncertain light of the lantern, she fancied that the colour was stealing back to Peter's face.

"He is alive; your master isn't dead!" she exclaimed to the hitherto apathetic Mahmed.

The announcement had an electrical effect upon the Indian boy. Peter dead was nothing to him; Peter living was his master for whom he had undoubted affection and devotion.

He began chafing Mostyn's hands, while Olive, now deadly tired, doggedly continued her efforts.

Mostyn's heart was now beating. His nostrils were quivering. He was breathing faintly, but with steadily increasing strength. Though partially choked by the water he had involuntarily swallowed when carried down by the ship, he had been saved from suffocation by his lifebelt, which kept his head clear of the water after he had regained the surface.

Restoring the circulation was the next step. Fortunately both the water and air were warm, and the dangerous consequences of a prolonged immersion were mitigated. Had the disaster occurred in other than tropical waters, the comparatively low temperature would have been fatal.

At length Peter opened his eyes. He was quite at a loss to grasp the situation. The lamplight puzzled him. At first he was under the impression that he was in his bunk, and that either Watcher Partridge or Watcher Plover had roused him to take in a signal. Somehow that didn't seem correct. Awkwardly he fumbled for the edge of the bunk board. Instead, his fingers encountered the stern-grating. Then his attention was wonderingly attracted by one of the knees of the after thwart. It had been split, and the sight of it irritated him, although he didn't know why, exactly.

He was beginning to realize that he was in a boat. How he got there, and why he should be in it, was a perplexity. It might be the Old Man's motor-launch—but no! Something was wrong somewhere.

A dozen fantastic theories flashed across his mind, only to be dismissed so unsatisfactorily that the failure made him angry. One thing he was certain of. Miss Baird was with him, but what she was doing there was a baffling problem. He wanted to speak to her, but hesitated lest that certainty should turn out to be an unreality.

He was still cudgelling his brain when he fell into a fitful and uneasy sleep.

The short tropical dawn was breaking when Peter awoke. He was now fully conscious of the events leading up to the foundering of the West Barbican, but was still at a loss to account for his presence in the boat. Stranger still it was to find that he had not been labouring under a hallucination with regard to Olive Baird.

The girl was sleeping on the bottom-boards, her head pillowed on a lifebelt. On the next thwart sat Mrs. Shallop, looking extremely dishevelled, with her black hair streaming in the wind. For once she was silent. On recovering consciousness she had grumbled considerably. Now there was no one to listen to her complaints. Peter had been asleep; Olive was still slumbering. Preston, although awake, was decidedly light-headed. As for

Mahmed and the two lascars, they were huddled together in the bows awaiting the appearance of the sun with its beneficent warmth.

Peter sat up wonderingly. His head swam a little, and he felt as weak as the proverbial kitten. Some one had covered him with an oilskin. He wondered who?

It came as a nasty shock to see poor old Preston stretched alongside, with one half of his face looking as if it had been battered in. The Acting Chief looked at Peter, but there was no recognition in the look.

"Hello, old man!" exclaimed Mostyn. "How goes it?"

The greeting was ignored. Preston made an effort to place his hand on his head. The attempt failed. With a groan the Acting Chief rolled over on his side.

"Water!" he gasped feebly.

Peter dragged the beaker from under the stern bench and moistened the injured man's lips. His own throat felt dry and parched, but already he realized the absolute necessity for husbanding the precious fluid.

Preston sighed and closed his eyes. For the time being Peter could do nothing more for the badly injured Acting Chief.

The Wireless Officer was feeling far too "groggy on his pins" to stand. Supporting himself by the gunwale, he knelt up and scanned the horizon. The wind was fresh and the sea fairly high, though regular. The boat, not under control, was driving broadside on to the wind, her high freeboard and comparatively light load allowing her to scud at quite a steady rate. Also, owing to the same circumstances, she rode the seas well, only an occasional flick of spray finding its way inboard.

The rain had ceased during the night, but the bottom-boards were awash. The masts and sails were still rolled up and stowed in a painted canvas cover. Beside them was a bundle of oars, and on top of them a rudder.

The fact that the boat was not under control stirred Peter to action. Having made sure that none of the rest of the *West Barbican's* boats was in sight, he aroused the inert lascars.

"Hai! hai!" he shouted. "Aft, you hands, and set sail."

The men showed no great haste to execute the command.

"Where go? India?" asked one.

"Lay aft, both of you," exclaimed Peter sternly, although in his weak state he found himself asking how he could enforce obedience. He knew enough of the native temperament to realize that if he gave a command and failed to see it carried out, his authority over the lascars was as good as gone for ever.

"Me tired," objected the other. "No *pani*, no *padi*."

Without another word Mostyn produced and ostentatiously displayed his automatic. There were great odds against its efficacy, after being submerged for several hours. The cartridges were supposed to be watertight, and were well greased. He had little fear on that score. The difficulty lay in the fact that the delicate mechanism of the pistol might have been deranged through the action of the salt water.

He felt confident that he could rely upon Mahmed. The boy was a devoted servant, and true to his salt. And Peter had no doubt about Miss Baird's ability to aid him if the lascars proved openly mutinous. For the present Preston was out of the running, while Mrs. Shallop was literally and figuratively a "passenger".

Greatly to Mostyn's relief the sight of the automatic acted like an electric shock upon the two lascars. With great agility and speed they began casting off the sail-cover and setting up the heavy mast.

While they were hoisting the lug-sail Mahmed shipped the rudder, and soon the boat was slipping along before the breeze.

Peter had been puzzling over the course for some considerable time. Against the westerly breeze he knew that days might elapse before the boat made the Mozambique coast. Being light and not provided with a centre-board, she was unable to sail at all close to the wind. In fact, it was doubtful whether she would make to windward at all. On the other hand, she would run well, and, with the knowledge that the island of Madagascar was somewhere under his lee — it might be anything between two hundred and four hundred miles — Mostyn decided that the best chance lay in making for it. There was, of course, a great possibility of several vessels being in the vicinity. If the boat were sighted, so much the better. If not — well, they would have to "stick it out" on very short rations.

A thorough search in the after locker disclosed the fact that there was an airtight tin containing fourteen pounds of biscuits, another lantern and a pound of tallow candles, a lead-line, some rusty fishing hooks and lines (relics of a long-forgotten fishing expedition), a hatchet, grass rope, and half a dozen signal rockets. Elsewhere in the boat were a small compass, a water-beaker about three-quarters full, spare oars, baler, boat-hook, grapnel, and a jib and mizzen sails, besides the lug that had already been set.

The baler had been nearly filled with rain-water during the night, but the lascars had drunk every drop. Peter, of course, was ignorant of this, and when he served out a small quantity all round the lascars must have congratulated themselves on their astuteness.

The tin of biscuits was then broached, and one biscuit handed to each person in the boat. Preston munched his ravenously, although every movement of his jaw caused him intense agony. He was still lightheaded, muttering incoherently about taking over the middle watch.

Olive was hungry and ate the "hard tack" with zest, but Mrs. Shallop pettishly declined her share as being unfit for a lady to eat. She even began her now well-known speech of self-advertisement, when Peter cut her short.

"I can offer you nothing better," he said curtly. "I would advise you to keep it, because you'll want it badly before long. And please understand there must be no grumbling. It has a bad effect upon the lascars."

"Surely I can talk if I want to?" protested the woman.

"Within limits, yes," replied Mostyn. "But I would point out that it would be far better if you did something useful. There's Preston, for instance, he requires pretty constant attention."

"Oh, Miss Baird can see to him," declared Mrs. Shallop. "She's younger than I am."

"Considering Miss Baird had three cases on her hands during the night—you, Preston, and myself—I think she's done more than her fair share," said Peter, and, filled with disgust, he turned to the helm, which Mahmed had temporarily taken.

He could see Olive's face flush under the selfish rudeness of the parvenue, but the girl, repressing her impulse to reply heatedly, remained silent.

A stiff glass of brandy, and the sound sleep resulting from it, had kept Mrs. Shallop in ignorance of her narrow escape from death in the disaster to the *West Barbican*. She was in the habit of consuming the contents of a bottle of strong waters per week. "By Dr. Selwyn's orders," she would explain. "He says I must have it, and it must be the very best." And Selwyn was never more astonished than when he heard of the prescription that was attributed to him. When the ship shook under the explosion a steward had rushed to Mrs. Shallop's cabin, and, unceremoniously dragging that lady from her bunk, had carried her along the alleyway to the companion ladder. Here the lady promptly collapsed. Meanwhile Mr. Shallop, who had been in the smoking-room, had gone on deck. In the darkness he saw nothing of his wife, and concluded that she was amongst the first to get away in the boats. At which he congratulated himself. He was spared the ordeal of being cooped up with Mrs. Shallop, who would to a certainty vent her anger upon him for having taken the sea voyage, although it was entirely on her suggestion that the ill-assorted couple booked passages on the S.S. *West Barbican.*

"This isn't going to be a picnic, I can see," soliloquized Peter, as he glanced to wind'ard. "It's up to me to do something now. I wonder if the Old Man would have logged me for this? Decent old chap, Bullock. I suppose he's gone."

Mostyn was steering due east by compass. He had no idea of the magnetic variation in this part of the Indian Ocean, neither had he any knowledge of the deviation of that particular compass. By steering due east he was hoping to effect a landing between the north and south of Madagascar—a fairly generous target of 1000 miles in length.

It was responsibility with a vengeance. Not only had the Wireless Officer to take over executive duties; he had to navigate the boat, regulate the supply of food and water, and maintain discipline until such times as Preston recovered and was able to take command. Judging by the injured man's appearance that day was still very remote.

Meanwhile Peter Mostyn, hiked by fate into the joys and difficulties of command, accepted the situation with typical British grit.

"I'll just carry on and make the best of it," he decided. "It won't be for want of trying if I don't get the boat safely to shore."

CHAPTER XXIV
Tidings from the Sea

"It's about time we had a letter from Peter from Cape Town, isn't it?" inquired Mrs. Mostyn.

Captain Mostyn deliberately lighted a cigarette while he worked out a mental sum.

"Hardly," he replied. "Give the Royal Mail a chance, old lady. We heard from the boy from Las Palmas. That ought to keep you satisfied for another week or so. By that time we ought to see the announcement of the *West Barbican's* arrival at Pangawani. Let me see: it was ten days ago when we saw the news of her departure from Durban. By Jove, old lady, we'll have a jollification when we know that the steelwork is handed over to the Kilba Protectorate Government."

There was no doubt about it. Captain Mostyn was worrying over the contract. The actual manufacturing of the bridge material had caused him very little anxiety. The keenness with which he had followed the work, the personal attention he gave to all the details, and the professional supervision of the whole process of manufacture had kept him busy both mentally and physically. But from the time the consignment was shipped on board the *West Barbican* at Brocklington he was metaphorically on pins and needles.

The contract was to include delivery at Pangawani. There were certain risks in the long sea passage that were to be taken into account. Unavoidable accidents might occur, that the most skilful master in the Merchant Service could not avert. Pangawani Harbour, with its shifting bar, had a sinister reputation in insurance company circles. That fact had resulted in the refusal of every underwriter whom Captain Mostyn approached to insure the steelwork to anything like its full value. The best terms he could obtain were 75 per cent, while the *West Barbican* was between the United Kingdom and Table Bay, and 66-2/3 per cent between Table Bay and Pangawani. That meant the bankruptcy of the Brocklington Ironworks Company should the steelwork fail to reach its destination, since every pound of available capital had been sunk in Captain Mostyn's "great push".

Curiously enough, his anxiety was solely for the safety of the steelwork. The knowledge that his son was on the very boat that was taking out the consignment hardly entered into his calculations. An indescribable faith in Peter caused him to regard the lad as being well able to take care of himself, happen what might. The ship might be lost, but Peter would be sure to come out all right.

Captain Mostyn and his wife were still discussing the movements of the *West Barbican*, and speculating upon the date of her arrival at Pangawani, when one of the maids brought in the evening paper, which was regularly left at the house by a newsboy from the village.

The Captain's first consideration was given to the Shipping List. The *West Barbican* did not appear.

"I told you so, my dear," he remarked. "We'll have to wait a little longer. Let me see; you want the serial page. Here you are."

Peter's father, always methodical, took a paper-knife from the writing-bureau and carefully cut the newspaper in half. Handing the back page to his wife, he settled down to read the news, notwithstanding the fact that most of it was reproduced from the London dailies, which he had already digested early that morning.

Mrs. Mostyn settled down for a comfortable evening. The fire was burning brightly in the open well-grate, the arm-chair was most comfortable. With the serial page and a half-finished jumper to work at while she read, Mrs. Mostyn meant to have a quiet and restful evening's amusement.

Presently she finished the instalment of the serial. She hardly knew what to think of it. Its abrupt ending made her angry with the author, or whoever was responsible for the conclusion, while the thrilling curtain left her on thorns as to what was going to happen in the next instalment. The rest of the page usually contained very little of feminine interest, consisting mainly of sporting topics and lurid testimonials to so-and-so's patent medicines.

Quite casually her eye caught sight of a badly printed paragraph in the Stop Press column. She read it through without the full significance of it coming home to her. Then she re-read it slowly and haltingly, as if every word was burning into her brain.

"John!" she exclaimed.

"Half a moment, my dear," protested Captain Mostyn, deep in an article dealing with the coal industry.

"John!" she said again.

Captain Mostyn glanced over the top of his half of the paper. He did not like being disturbed. It usually meant that his wife had discovered a stupendous bargain in the sales column, with the inevitable result.

"Good Heavens, old lady!" he ejaculated, greatly alarmed at the grey, drawn expression on his wife's face. "What is it?"

Mrs. Mostyn did not reply. With trembling hands she gave the paper to her husband, and pointed to the grim announcement in the Stop Press column:

"Lloyd's agent at East London telegraphs, 'S.S. *Maréchal Foch* arrived here to-day with eighteen lascars, survivors of the S.S. *West Barbican*, which foundered in the Mozambique Channel on the night of the 22nd. No trace has been found of the ship's officers and the remainder of the crew. Survivors cannot give any explanation of how the disaster occurred.'"

"Peter!" gasped Mrs. Mostyn.

Her husband was thunderstruck. The gravity of the news had taken him completely aback. He gave no thought to the precious steelwork. His whole concern was for his son.

The bald announcement was serious enough in all conscience. Reading between the lines it gave scant hope that there might be other survivors. Was it possible that Peter had in his prime fallen a victim to the remorseless sea?

"There's nothing very definite, my dear," he remarked as calmly as he could. "Perhaps to-morrow we'll hear that some more boats have been picked up. Strange things happen at sea."

Mrs. Mostyn shook her head. After Peter's almost miraculous return when given up for dead, after the S.S. *Donibristle* had been reported "overdue, missing, and believed a total loss", she could hardly hope for a second intervention of Providence.

"Tut, tut," said Captain Mostyn, his forced manner belying the doubts that assailed him. "Why shouldn't he turn up trumps a second time? Why, I know an old pensioner at Portsmouth who, during his twenty-one years' sea life, was reported killed four times. And he's hale and hearty to-day at eighty-five, or he was when I heard of him a fortnight ago. I'll see my friend Parsons at Lloyd's to-morrow. He'll keep me posted as to the latest news. Peter will be all right, never fear."

But Captain Mostyn had his doubts. He knew enough about the sea to realize the possibility of his son going down with the ship. He argued that the disaster must have been sudden, since there was no mention of the ill-fated *West Barbican* having sent out wireless messages for aid. That pointed to the vessel foundering in a few minutes; in which case there had not been time to lower all the boats. Quite likely the one containing the eighteen lascars was the only one successfully lowered. Again, the absence of an officer in the boat pointed to a complete disorganization of discipline. On the face of Lloyd's telegraphed report things looked very black indeed.

Captain Mostyn spent a sleepless night, but he hardly gave another thought to his financial losses. Over and over again he tried to reconstruct the scene on board the sinking liner, with the object of convincing himself that his son had escaped with his life. Throughout the long night he was building up suggestions and immediately demolishing them on account of an incontestable flaw in the theory.

Next day Captain Mostyn went up to town by his usual train, but, instead of proceeding to the offices of the Brocklington Ironworks Company, he went straight to Lloyd's. Here he was informed that no further news of the loss of the S.S. *West Barbican* had been received, but the detailed report of the Master of the S.S. *Maréchal Foch* was expected by cable that day.

The same afternoon there was a hurriedly convened meeting of the directors of the Company. None of them had noticed the announcement concerning the *West Barbican* in the papers, and Captain Mostyn's bald statement came as a complete surprise. No definite steps could be taken until the ship was officially reported lost, and then only would the underwriters pay the 66-2/3 per cent of the value of the steel-work.

A fortnight or more passed, with nothing to break the silence that seemed to be brooding over the loss of the *West Barbican*. For some reason the report of the captain of the *Maréchal Foch* had not materialized. It afterwards transpired that he was in hospital at East London.

At last the silence was broken by the receipt of a Press Association cablegram from Port Louis, Mauritius:

"Portuguese sailing ship *Balsamao*, Lorenzo Marques to Goa, arrived here to-day with sixteen Europeans and eleven Indians, survivors of the S.S. *West Barbican*. Names of the Europeans as follows: Anstey, Crawford, M'Gee, Peterson, Fulwood, Selwyn, Wright, Scott, Palmer, Partridge, Plover, Smith, Fostin, Applegarth, and Shallop (passenger)."

A ray of hope flashed across the minds of Peter's parents. The name "Fostin": it was possible that it was a telegraphic error for "Mostyn". The conviction grew until Captain and Mrs. Mostyn felt perfectly convinced that the name in question was actually supposed to represent that of their son.

But, alas! disillusionment came next day when Captain Mostyn paid a visit to the offices of the Blue Crescent Line, and was given a list of the names of the officers and crew of the ill-starred *West Barbican*. Amongst them was: "Geo. Fostin, steward".

"We are afraid to have to admit that Captain Bullock is amongst the missing," said the secretary of the Blue Crescent Line to Captain Mostyn. "One of our senior and most experienced skippers, and on his last voyage before retiring. The Chief Officer, Mr. Preston, is also missing, also the Wireless Officer. It can only be surmised that they stuck to the ship to the last and went down with her. The Wireless Officer's name is—let me see."

The official referred to the list in front of him.

"The same as yours, sir," he continued. "A relation, perhaps?"

"My son," replied Captain Mostyn sadly yet proudly.

CHAPTER XXV
Riding it Out

"What is the time, please, Miss Baird?" inquired Peter.

"Nine o'clock," replied Olive, consulting her wristlet watch, the only one of five in the boat that had survived.

"Too early for grub, then," continued Mostyn "We must economize. And with water, too. It's going to be a scorching hot day."

He omitted to add that in all probability there would be a stiffish wind before long, possibly increasing to hurricane force. The thundery rain, coming before the wind, pointed to a severe blow before many hours were past. Meanwhile the breeze had dropped until the boat was making less than one knot.

Peter had practically shaken off the effects of his prolonged immersion. He was feeling a bit stiff in the limbs, and had developed a healthy hunger. The latter troubled him far more than the stiffness. Work would relieve his cramped arms, but it would also increase the pangs of the inner man.

In the light breeze he could safely entrust the helm to one of the lascars, provided he kept his weather eye lifting in case a sudden squall swept down upon the boat. The native might or might not be able to handle a sailing craft, but Peter was resolved to take no risks on that score. He would rather place Olive at the helm, although in the event of danger he meant to stick to the tiller for hours if needs be.

"Due east, *mutli*," ordered Mostyn, having signed to the lascar to come aft.

The man nodded and repeated the compass course. Since Peter had displayed his automatic the pair of lascars had been remarkably tractable.

The Wireless Officer's next step was to rig up a tent to shelter the women from the blazing sun. Calling Mahmed to assist him, he lashed the unshipped mizzen mast to the mainmast just below the goose-neck of the latter, so that the boom could swing out in the event of a gybe without fouling the almost horizontal ridge-pole. The after end of the mizzen was propped up by a crutch made by lashing a couple of boat-stretchers crosswise. Over this was spread the mizzen sail, the ends of the ridge-tent being enclosed by means of the jib and a couple of oilskin jackets.

"There you are," declared Peter, surveying the result of the joint handiwork of Mahmed and himself. "You'll be sheltered under the sail. I would advise you both to sleep during the heat of the day."

Olive declined, with a smile, adding that she preferred to be in the open air. Mrs. Shallop hardly deigned to acknowledge the effort Mostyn had made for her comfort as far as lay in the resources at his command.

She had not been under the tent for more than a minute, when she reappeared holding up a ring-bedecked hand for inspection.

"I've lost a diamond out of this ring," she announced in a loud voice; "and it's a valuable one. It cost a sovereign."

Peter could not help smiling.

"Whatever can one do with a female like that?" he soliloquized. "The loss of a twopenny-halfpenny stone is of more consequence to her than the chance of losing her life."

Contriving to conceal his amusement he replied: "It can't have gone very far, Mrs. Shallop, if you had it in the boat. We'll probably find it under the bottom-boards."

"Then make those blacks look at once," ordered the lady peremptorily.

Peter pretended not to have heard the strident, imperious command. It would have been waste of breath to point out that the boat could not be searched without disturbing Preston, and that the awkwardly placed bottom-boards could not be removed while the boat was under way.

With a parting shot at the young officer for his incivility, Mrs. Shallop retired to the tent and began to nag Miss Baird, who had shown no disposition to assist in the search.

"Thanks, Mr. Mostyn," said the girl, when Peter warned her of the heat of the sun. "I'm quite all right. You see, I took the precaution of wearing a topee when we were ordered into the boat. May I steer?"

For a second time that morning Mostyn relinquished the helm. Then, having seen that Preston was as comfortable as possible, he sat on one of the side-benches and chatted to the helmswoman. Even then he was not idle, for, on the principle that "you never know when it may be wanted", he took his automatic pistol to pieces and carefully cleaned the mechanism, sparingly oiling the working parts with a few drops of oil from the lamp.

"Do you know how this thing works?" inquired Peter casually.

"Yes," replied the girl promptly. "You have to pull back the hammer for the first shot, and as long as the trigger is pressed the pistol goes on firing until the magazine is empty."

"I wonder how you know," thought Mostyn.

He shook his head.

"This pistol doesn't," he explained. "Some simply act automatically as long as the trigger is pressed. That's rather a drawback if a fellow's a bit jumpy. He's apt to let fly a hail of bullets indiscriminately. No! This pistol of mine cocks itself after every shot, and it requires another pull on the trigger to fire each of the succeeding cartridges."

"The one I saw was different," rejoined the girl. "It was my brother's. He was killed at Ypres in '18."

Peter politely murmured regrets, but inwardly he felt relieved that the fellow who had instructed Olive into the mysteries of automatic pistols was only a brother.

Just then Preston roused slightly and asked for water.

"Better, old man?" asked Mostyn, as he poured a few precious drops into the baler, and held the rim to the Acting Chief's dry lips.

"Hocussed an' sandbagged, that's what's happened to me," mumbled Preston thickly. "Where the hooligan Harry am I?" And, with a sudden movement, he jerked the baler out of Peter's hand.

The man was obviously still delirious. Before Mostyn could decide what to reply, Preston shut his eyes and went to sleep again.

Mostyn picked up the baler from where it had fallen under the stern-bench. A couple of spoonfuls of fresh water had been wasted.

"Is that land?" suddenly inquired Olive, pointing away on the port bow, where a low, dark line was just visible on the horizon, looking very much like a chain of serrated mountains.

"Cloud bank," replied Peter briefly. Then in explanation he added: "There's wind behind that lot, Miss Baird; probably more than we want. It may head us too."

Glancing into the compass hood to see that the girl was steering a correct course, Mostyn rapped on the thwart immediately abaft the canvas shelter in which Mrs. Shallop was either resting or brooding over more or less imaginary grievances.

"We'll have to unrig the tent," he announced. "There's a stiff breeze bearing down on us."

"I don't like stiff breezes," retorted the lady promptly. "I'd rather have the tent up to keep the wind out."

"Sorry," replied the Wireless Officer. "It can't be done. In two minutes the lascars will commence unrigging the tent."

Mostyn allowed a good three minutes to elapse before signing to Mahmed and the lascars to take down the canvas. It was an absolutely necessary step, in order to allow unimpeded access to the working canvas, should it be required either to reef the sail or stow it altogether.

Having seen the task carried out, Peter proceeded to rig up a sea-anchor.

"It may come in jolly useful," he remarked to Miss Baird. "If we don't want it I won't complain about useless work."

With the assistance of the three Indians Mostyn bent a rope span to the yard and boom of the mizzen sail. Through the centre of each span he secured a stout grass warp, weighting the yard with the grapnel, so that, if it became necessary to ride to the improvised sea-anchor, the grapnel would keep the sail taut and in a vertical plane.

By the time these preparations were completed the bank of ragged-edged clouds had covered most of the sky to wind'ard. The sun was beginning to become obscured, while there was an appreciable drop in the temperature of the air. The wind had fallen away utterly, leaving the sail hanging idly from the yard. The water no longer rippled under the boat's forefoot. All was silent save for the creaking of the mast and spars as the boat rolled sluggishly in the long, gentle swell.

Keenly on the alert, Peter had taken over the helm, and was keeping a sharp look-out to wind'ard.

"Down sail!" he ordered.

The canvas was lowered and stowed. As a precautionary measure Mostyn had the sea-anchor hove overboard, trusting that at the first squall the high, freeboarded boat would drift rapidly until brought head to wind by the drag of the improvised floating breakwater.

"It's coming," said Olive in a low voice, as a long-drawn shriek could be faintly heard—the harbinger of a vicious squall.

By now it was almost dusk, so dense were the clouds overhead. The tropical sun had no power to penetrate the sombre masses of vapour. Less than half a mile to wind'ard the hitherto tranquil water was white with wind-lashed foam; while, in strange contrast, the sea-anchor was rubbing gently alongside the boat in the perfectly smooth sea.

Louder and louder grew the volume of sound, until with a vicious rush the squall swept down upon the boat. For a few seconds, while she lay broadside on, the boat heeled to such an extent, under the wind-pressure upon her high sides, that the water was actually pouring in over the lee gunwale. Then, spinning round as the grass rope attached to the sea-anchor tautened, the boat rode head to wind and sea.

In a brief space of time the terrific gusts had raised quite a mountainous sea, with deep troughs and short, sharp crests which, torn by the blasts into clouds of spindrift, flew completely over the boat. So far she had ridden it out splendidly, the sea-anchor breaking the more dangerous waves in a manner that was quite astonishing. Yet the while the grass rope was snubbing wickedly in spite of its natural springiness. Through the clouds of spray Peter could see that the lascars for'ard were betraying considerable uneasiness lest the rope should part.

Mostyn too realized the danger. He regretted that he had not doubled the rope, but now nothing could be done beyond putting a temporary "parcelling" round it where it passed through the bow fairlead.

More than once the Wireless Officer gave a hurried glance at Miss Baird. Outwardly the girl seemed perfectly self-possessed, and, with her natural thoughtfulness, she was sitting on the stern-gratings and doing her best to keep the still delirious Preston from sliding from side to side with the erratic and disconcerting motion of the boat.

The squall lasted for perhaps five minutes. Then, after a lull, came another series of vicious blasts from a different point, that was almost at right angles to the direction of the initial squall. This had the effect of raising a nasty cross-sea, accompanied by a torrential downpour of rain.

Suddenly, at less than a couple of cable-lengths to windward, appeared the misty outlines of a tramp steamer. She was labouring badly, rolling almost rail under and throwing up showers of spray high above her bridge.

Standing up and keeping his feet with difficulty Mostyn frantically waved to the vessel. Mahmed followed his example and also hailed in his high-pitched key. Shouting was useless. No volume of sound short of that of a fog-horn could possibly have carried that distance in the face of the howling elements.

The next instant the temporary clearing of the downpour gave place to a blinding deluge. The steamer vanished as utterly as if she had suddenly plunged to the bed of the ocean.

"Has she seen us?" inquired Olive, raising her voice.

"'Fraid not," replied Peter, still staring in the direction where he had last seen the tramp. "Couldn't do much if she did in this dust-up. I'll risk a rocket, any old way."

Some time elapsed before a rocket could be taken from its airtight case, and the touch-paper ignited. Then with a hiss the detonating signal soared obliquely upwards, its intended course deflected by the terrific wind.

It burst at less than a hundred feet in the air, but the report was so faint and the flash so weak that Mostyn could only reiterate his doubts as to whether the tramp could see or hear anything.

"It's lucky she didn't run us down," he added. "I know those blighters. They think they've got the whole ocean to themselves and carry on at full speed. In fog it's often the same, the idea being to get into better weather as soon as possible."

For another ten minutes it blew hard, but, thanks to the improvised sea-anchor, the boat was making very little leeway and riding head to wind. Occasionally the crested tops of the cross-seas flopped in over the gunwale, and the two lascars were kept baling steadily. Olive and Mahmed were tending the still delirious Preston, the former holding him to prevent further injuries to his badly damaged head, while the boy kept a strip of painted canvas over the Acting Chief's body to shelter him from the rain and spray. Mrs. Shallop was the only idler. Refusing Peter's offer of his oilskin, she sat huddled up on the bottom-boards, with the water swirling over her feet and her clothing saturated with the torrential rain. Too dispirited to use her voice in complaint, she sat and shivered in morose silence, posing as a martyr and yet getting no sympathy from anyone.

At length the wind ceased, although the rain continued in violence. This had the effect of calming the water considerably, and Peter took the opportunity of ordering the lascars to spread out the square of painted canvas, and catch as much rain as possible to augment the precious store of fresh water.

Within an hour the sky cleared and the wind freshened into a one-reef breeze. The sea-anchor was taken in and sail again set; but there was the disquieting knowledge that the wind was dead in their teeth. Either the boat must be kept "full and bye", gaining little or nothing on each tack, or Mostyn must "up helm" and retrace his course on the chance of making the now far-distant Mozambique shore, which meant that the previous sixteen-hour run was utterly wasted.

"If only we had a motor!" he exclaimed.

CHAPTER XXVI
Mostyn's Watch

Just before sunset the wind dropped to a flat calm. Peter took advantage of the practically motionless conditions to employ the fishing-lines that had been discovered in the after locker. The hooks were sharpened by means of the sandpaper fixed to the solitary box of matches in the boat. Small pieces of biscuit, soaked in water and rolled between the finger and thumb, served as bait. The lines were old and far from sound, but might be relied upon to bear a steady strain of about seven pounds.

"Do we fish on the bottom, Mr. Mostyn?" asked Olive facetiously.

"Yes, rather," replied Peter, entering into the jovial spirit. "That is, if your line is long enough. We're only about a mile from the nearest land, and that's immediately beneath us."

Olive lowered her line steadily. Before she had paid out half of it there was a perceptible jerk and the line slackened.

"I've struck soundings," she reported.

At first Mostyn thought that the girl was still joking, but an exclamation from one of the lascars, who was lowering one of the lines, convinced him that the lead weights had touched something of a solid nature.

Taking Miss Baird's line, Peter held it between his extended first and middle fingers. He could distinctly feel the lead trailing over a hard bottom, as the boat was carried along by a slight current.

"Strange," he ejaculated. "We're in less than five fathoms. I had no idea that there was a shoal hereabouts."

Steadying himself by the mast, Mostyn stood upon the gunwale and scanned the horizon. North, south, east, and west the aspect was much the same—an unbroken expanse of water, differing in colour according to the bearing. To the east it was sombre, to the west the sea was crimson, as it reflected the gorgeous tints of the setting sun.

"No land in sight," he reported.

The shoal proved to be a good fishing-ground, for, before the short tropical dusk had given place to night, a dozen fair-sized fish, somewhat resembling the herring of northern waters, had been hauled into the boat.

"What is the use of them after all?" inquired Olive. "We can't cook them, and raw fish are uneatable."

"Unpalatable, Miss Baird," corrected Peter. "It is just likely that we shall have to eat them. To-morrow we'll try curing them in the sun."

"Couldn't we fry them over the lamp?" asked the girl, who obviously had not taken kindly to the suggestion that the fish should be sun-cured. She was extremely practical on most points, but she drew the line at dried but otherwise raw herrings.

"You might try cooking for yourself, Miss Baird," said Peter dubiously. "You see, we have to economize in oil almost as much as with water; but I think we can stretch a point in your favour."

"In that case I'd rather not," rejoined the girl decidedly. "It wouldn't be fair to the rest, and there's the oil to be taken into consideration. I hadn't thought of that."

Having caught sufficient fish for their needs, the anglers hauled in their lines and stowed them away. Peter then shared out half a biscuit apiece and a small quantity of water. This time Mrs. Shallop was not too proud to accept the meagre fare. She ate her portion of biscuit, and even suggested to her companion that if Olive had more than she wanted she could give it to her.

Watches were then set for the night, Mahmed and one of the lascars taking from eight till two, and Peter and the other lascar from two till eight; the time being determined by Miss Baird's watch. This meant a long trick, but it was unavoidable. The three natives had been standing easy most of the day, while Peter had had no sound sleep for nearly thirty hours.

"I am not going to sleep in that tent, Mr. Mostyn," declared Olive, with an air of finality, speaking in a low voice. "I'd much rather curl up on the bottom-boards. It's not nearly so stuffy."

"Is it because Mrs. Shallop has been jawing?" asked Peter. "I'll tell you what; there's a square of spare canvas sufficient to rig you up a shelter between those two thwarts."

"Don't bother!" exclaimed Mrs. Shallop, who, when she wanted, was marvellously quick of hearing. "You can have the tent. I'll sleep outside."

And, before the astonished Peter and Olive could say anything, Mrs. Shallop snatched up the piece of canvas and went for'ard.

"She's ashamed of herself and is trying to make good, I think," suggested Mostyn. "Well, your pitch is queered, Miss Baird, but there's the tent."

Without a word Olive disappeared behind the flap.

Mostyn could rely upon Mahmed to keep his companion "up to scratch", so with an easy mind the Wireless Officer went for'ard, wrapped himself in his oilskin, and was soon sleeping soundly on the bottom-boards.

He was awakened by Mahmed at the stipulated hour. In his drowsiness it was some moments before he realized where he was, and it rather perplexed him to find his boy shaking him by the shoulder without the customary "Char, sahib".

It was a bright, starlit night. The wind was soft and steady, and the boat was rippling through the water at at least four knots.

Going aft, Mostyn peered at the compass. There was sufficient light to enable the helmsman to steer without having to use the candle-lamp of the binnacle. The course was still sou'-east, or four points south of the desired direction. It was as close as the boat could sail; even then she made a lot of leeway.

"Not'ing to report, sahib," declared Mahmed.

"All right," was the rejoinder. "Carry on."

The lascar told off to share Mostyn's watch came aft, rubbing his eyes and yawning.

"Me no well, sahib," he said. "Me tink me die."

"Take the wheel," ordered Peter, using the term instead of tiller, since the lascar was well acquainted with the word "wheel".

The man grasped the tiller without another word. His little ruse was a "wash-out", and, finding that his imaginary ailment received no sympathy, he carried on as if nothing had happened.

Mostyn then proceeded to attend to his injured brother-officer, washing his wounds and feeding him with biscuit.

Preston was still very weak, but quite rational in his speech. His prolonged sleep had restored his mental powers, but he was unable to move without assistance.

"What's happened, old man?" he inquired. "I've been racking my brains to find out how I got laid out. I remember lowering away the boat, and after that everything's a blank."

"You got a smack with the lower block swaying," replied Peter. "At least that's what I was told. They didn't pick me up for a couple of hours or more after the ship went down."

"And the Old Man?" asked Preston.

The Wireless Officer shook his head sadly.

"'Fraid he's done in," he answered. "When the ship disappeared he was with me on the bridge. I never set eyes on him after that."

"Rough luck," murmured Preston. "His last voyage before he went on the beach with a pension. Sound old chap too, although hard to get on with at times."

"One of the best," declared Mostyn.

There was silence for a few moments.

"Mostyn, old son," exclaimed Preston. "How about a cigarette?"

"Wish I could oblige you," replied Peter; "but there isn't a shred of tobacco in the boat. I had my case full in the wireless-room when she sank—a silver presentation case—and I quite forgot to ram it into my pocket."

The Acting Chief smiled wanly, and immediately regretted having done so. It was a painful process, with one side of his face battered.

"You ought to have known better than that," he remarked reprovingly. "Especially as you've been through much the same sort of thing before. Tobacco takes the edge off a fellow's hunger. I suppose your case was watertight?"

"I think so," replied Peter. "But since I haven't got it I don't see that it matters."

"Mostyn, dear old thing, you don't deserve pity," said Preston. "Just feel in the inside pocket of my coat. Luckily I haven't been in the ditch."

Peter did as requested, and drew out a cardboard box containing nearly a hundred Virginias.

"Lifted 'em from the Chief Steward's cabin," explained the Acting Chief. "They would have gone to Davy Jones if I hadn't. Take charge of them, old man. They'll last the pair of us for a fortnight, and by that time——"

"How about the lascars?" asked Pater.

"Mohammedans," rejoined Preston briefly. "They aren't allowed to smoke. At least," he added, "I don't think they do. Of course, they'll come in if they want any. We'll see. Light up for me, old fellow."

"We collared a box of matches from you," said Peter. "These are all we have on board. They are yours, of course, but— —"

"Do they strike?" asked the Acting Chief. "I've had them for at least a twelvemonth. Sort of emergency issue, don't you know. Try my pockets, old son. I've a lighter somewhere, I'll stake my affidavit on that— — Gently, old man!"

"Sorry," exclaimed the Wireless Officer. "By Jove, Preston, you are a marvel."

"Rot!" ejaculated the other in self-depreciation. "Merely a case of looking after one's own interests."

Placing the end of a cigarette between Preston's lips Peter lit it. The Acting Chief grunted contentedly.

"There's a box of Turkish delight in my pocket," he continued. "Take it and hand it to the womenfolk. All the joy hasn't gone out of life yet, Sparks. Light up and get happy."

Mostyn did so. Never before had he so appreciated the soothing effect of a cigarette.

In this complaisant state of mind he was addressed by the lascar at the helm.

"Mahometan smoke, Sahib; Sikh, Mahometan, too: him not smoke."

Which resulted in the tip of another cigarette glowing in the darkness.

"I feel a jolly sight better for that," declared Preston gratefully, when the cigarette was finished. "Think I'll have another caulk. S'pose you don't mind?"

"Not at all," replied Peter. "Carry on. It will do you good. Are your bandages comfortable?"

In a few minutes the Acting Chief was slumbering more peacefully than he had done since his accident. Mostyn, left to commune with his own thoughts, squatted on the weather side of the stern-sheets so that he could give an occasional glance at the compass, and keep an eye on the lascar at the tiller.

It was a long trick. It seemed as if the eastern sky would never pale to herald the dawn of another day.

At 4 a.m. the boat was put on the starboard tack, the wind still heading her as before. Then, having trimmed sheets, Mostyn took the tiller and ordered the lascar into the bows.

At length the dawn broke—not a pale grey, as Peter had hoped for, but with far-flung lances of vivid scarlet. That indicated rain and wind before the day was done.

There was a movement of the canvas awning, and, somewhat to Peter's surprise, Miss Baird emerged cautiously, crawling, since there was no other means of negotiating the narrow gap that served as a door.

She was bareheaded, her hair trailing over her shoulders in two long plaits. The outward and visible signs of her costume consisted of a yellow oilskin. Silhouetted against the red glow of the sky she looked as if she were outlined in deep gold.

"Good morning, Miss Baird," observed Peter politely. "You're out early."

"I simply couldn't sleep any longer," replied the girl. "I hope you don't mind my intruding upon you? What a glorious sunrise."

"From an artistic point, yes," agreed Mostyn. "But I'm afraid we'll get it before very long."

"She's a safe boat," said Olive with conviction. "She isn't exactly a yacht, but, personally, I'm rather enjoying it."

"Even on short rations?" inquired Peter.

"Up to the present, yes," was the reply. "It's rather a novelty being served out with biscuits, but I'm not looking forward to the sun-dried herrings."

"Perhaps," said Peter, producing the box of Turkish delight, "these will prove a welcome substitute for the herrings. No, don't thank me. Preston's the fellow."

With her eyes sparkling, Olive proceeded to count the luscious squares. Mostyn looked on, wondering at the reason of her act.

"Sixty-three, sixty-four," concluded the girl. "That's thirty-two for Mrs. Shallop. You'll be witness, Mr. Mostyn, that it's a fair divide?"

The Wireless Officer had said nothing about sharing the sweetmeats. Olive's generosity and fairness were all the more apparent.

"I'm out of a post, Mr. Mostyn," she continued, with a light-hearted laugh. "Mrs. Shallop and I are not on speaking terms."

"That rather gives you a free hand. I'm very glad," said Peter gravely.

"Yes," admitted the girl. "She has certainly been a bit trying of late. Do have a piece of Turkish delight?"

Mostyn shook his head.

"No, thanks," he declined. "Your share won't go very far. Besides, I'm in luck too. Preston had a big box of cigarettes in his pocket. So you're pleased to be free of Mrs. Shallop?"

"Rather," replied the girl whole-heartedly. "The only thing that troubles me is how I am to get home again, if we come through this adventure safely."

"Don't worry about that, Miss Baird," declared Peter boldly. "I'll see you safely home. You can be quite independent of that woman."

"Thank you so much," said Olive gratefully, and almost unconsciously she laid her hand lightly upon Peter's arm.

A thrill of pleasure swept across the Wireless Officer's mind. Then, as if to seal the compact, the tropical sun in all its glory appeared above the rim of the horizon.

"I'm not a woman," exclaimed a strident voice from inside the tent. "I'm a lady. I am really. My father was a naval officer—a captain."

The man and the girl looked at each other. Olive's face was wreathed in smiles. Peter actually winked. In the Eden that he had created the presence of the Serpent was of no account.

CHAPTER XXVII
Aground

The rest of the day until four in the afternoon passed almost uneventfully. The breeze still held, but blew steadily from the same quarter with hardly a point difference in eighteen hours. With one reef in the mainsail the boat had all she could carry with comfort, and, save for an occasional fleck of foam over the weather bow, was dry and fairly fast.

The disconcerting doubt in Peter's mind was whether the boat was making good to wind'ard. Apparently she was, but whether the leeway counter-balanced the distance made good, or whether the boat was actually losing on each tack remained at present an insolvable problem.

During the greater part of the day the heat of the sun was tempered by the cool breeze, but late in the afternoon more indigo-coloured clouds began to bank up to the east'ard. The roseate hues of early morn were about to vindicate themselves as harbingers of boisterous weather.

"Sea-anchor again, I suppose," soliloquized the skipper of the boat. "Beat and beat and beat again, then drift to lee'ard all we've made. We'll fetch somewhere some day, I expect."

He rather blamed himself for not having put the helm up directly the previous gale had blown itself out. Running before the easterly breeze would have brought the boat within sight of the Mozambique coast before now. On the other hand, how was he to know that the easterly breeze would hold for so many hours? It rarely did.

"It's a gamble," he thought philosophically. "I've backed the wrong horse. I've got to see this business through."

Once more the tent was struck. This time Mrs. Shallop, who had taken possession when Olive came out, made no audible protest. Possibly she was too busy eating Turkish delight. In that respect she acted upon the principle of "Never leave till to-morrow what you can eat to-day".

The sea-anchor was prepared ready to heave overboard. Loose gear was secured, and the baler placed in a convenient spot to commence operations should a particularly vicious sea break into the boat.

Darkness set in. No stars were visible to mitigate the intense blackness of the night. The candle-lamp of the boat-compass had to be lighted in order to enable the helmsman to keep the craft on her course. Its feeble rays faintly illuminated Peter's face as he steered. Beyond that it was impossible to distinguish anybody or anything in the boat, the bows of which were faintly silhouetted against the ghostly phosphorescence of the foam thrown aside by the stem.

So far there was no necessity to ride to the sea-anchor. The wind, slightly increasing in force, demanded another reef in the mainsail. No doubt the boat would have stood a whole mainsail, but Peter was too cautious and experienced to risk "cracking on" in a lightly trimmed craft unprovided with a centreboard or even a false keel.

The two lascars were told off to tend the halliards, Mahmed stood by the mainsheet, while Peter steered. The latter, his senses keenly on the alert, was listening intently for the unmistakable shriek that presages the sweeping down of a squall. In the utter darkness the sense of hearing was the only means of guarding against being surprised by a violent and overpowering blast of wind.

"It may not be so bad after all," he remarked to Olive, who had insisted on keeping by him at the tiller. "There's rain. I expected it. Luckily it's after the wind, so the chances are we've seen the worst of it."

It was now nearly ten o'clock. The boat had been footing it strongly, since Peter had eased her off a point. The seas were high—so high that between the crests the boat was momentarily becalmed. Yet, thanks to Mostyn's helmsmanship, she carried way splendidly, until the ascent of the on-coming crest enabled the wind-starved canvas to fill out again.

Very soon the few heavy drops gave place to the typical tropical downpour. Even had it been daylight it would have been a matter of difficulty to see a boat's length ahead. In the darkness it seemed like crouching under a waterfall. Breathing resulted in swallowing mouthfuls of moisture-laden air. In less than half a minute from the commencement of the downpour, there was an inch or more of water over the bottom-boards in spite of Mahmed's strenuous work with the baler.

Contrary to Peter's expectations, the strength of the wind did not appreciably diminish, but the rain had the effect of considerably beating down the crests of the waves.

It was now quite impossible to hear anything beyond the heavy patter of the big raindrops upon the boat. It was a continuous tattoo that outvied the roar of the wind. At this juncture the candle of the binnacle lamp blew out. To attempt to relight it was out of the question. Every part of the boat's interior was subject to a furious eddy of wind. A match would not burn a moment.

"Hardly good enough," decided Peter, wiping the moisture from his eyes. "I'll get canvas stowed and out sea-anchor till the worst of this is over."

With his disengaged hand Mostyn tapped Mahmed on the shoulder. Desisting from his task of baling, the boy looked into his master's face.

"Tell them to stow canvas," shouted Peter, indicating the invisible lascars crouching against the main thwart. "I'll tend the mainsheet. Look sharp!"

Mahmed raised himself and began to crawl over the thwarts on his way for'ard.

Suddenly there was a terrific shock. The boat seemed to jump a couple or three feet vertically, and then come to an abrupt stop with a jar that flung Peter from the tiller, and pitched Mahmed headlong until he was brought up by his head coming into contact with Mrs. Shallop's portly back. Olive, taken unawares, was jerked in a for'ard direction, until she saved herself from violent contact with stroke-bench by grasping Peter's arm. The pair subsided upon the gratings, narrowly missing what might have been a serious collision with the helpless Preston.

Mostyn regained his feet in double quick time, and made a grab at the tiller. The boat was aground, lifting to every wave that surged against her port-bow. That she was badly damaged there could be no doubt, since water was pouring in through a strained garboard.

Steadying himself by the now useless tiller, Peter peered anxiously into the darkness. Except for the phosphorescence of the breaking water alongside, there was nothing distinguishable. Sea and sky were blended into a uniform and impenetrable darkness.

Everyone on board the boat, although fully aware of the immediate danger, maintained silence. The grinding of the boat's planking upon the sharp rocks, the howling of the wind, and the swish of the breaking waves were the only audible sounds.

It seemed to Mostyn that, in his self-assumed position of skipper of the boat, he must do or say something. He did neither. He could form no sentence of encouragement; he was unable to take any action to further safeguard the lives and interests of his companions. He felt cool and collected, yet he had a suspicion that he "had the wind up". Try as he would his benumbed brain would not answer to his efforts.

It was Preston who broke the spell. Lying half-submerged in water, the Acting Chief was taking things calmly in spite of his physical disability.

"Sparks, old man," he exclaimed, "you look like losing your ticket. I do believe you've run us aground."

The silence was broken. Peter laughed at his companion's quip.

"We were making for land," he replied, "and now we've jolly well found it. Get out the rockets, Mahmed."

Mahmed had delivered Mostyn's order to the lascars. Already the sail had been hastily lowered. Its folds served as a screen to break the force of the wind, nevertheless, it was a difficult matter to keep a match alight sufficiently long to ignite the touch-paper of the rocket.

"Cheap and false economy, these things," thought Peter, as he wasted three matches in a vain attempt to kindle the touch-paper. "Why didn't the owners supply Verey pistols to all the boats?"

At length the fuse began to sizzle. An anxious fifteen seconds ensued. More than once the minute sparks looked as though they had given out, only to reappear with a healthier glow.

Then with a swish the rocket soared skywards, although with an erratic movement as it was caught and tossed about by the wind.

Mostyn made no attempt to follow its course with his eyes. Holding a hand to his brows he gazed in the direction in which he expected to see land.

A vivid glare overhead, as the rocket threw out a series of blue star-shells, revealed what he wanted to know. Eighty or a hundred yards ahead was a line of cliff, fronted by a gently shelving stretch of sand. The boat had struck on the apex of a reef. She was neither on a lee nor a weather shore, but rather on the dividing line of each.

"Good enough," shouted Peter encouragingly. "Light the lantern, Mahmed."

The boy succeeded in getting the lamp alight. Even its feeble glimmer put a different complexion upon things.

Beckoning the lascars aft, Mostyn sent one of them back again to bend the warp to the anchor and throw the latter overboard, in case the badly damaged boat should be washed off the reef.

This done, the question arose: how were the women and Preston to be taken ashore?

"Take Mr. Preston," said Olive. "I can walk."

"Easy enough if it's shoal water right up to the beach, Miss Baird," rejoined Peter, "That we'll have to find out. I think I'll rope you together."

Preparations for abandoning the boat having been completed, Peter led the way, holding aloft the lantern. Behind him came the two lascars, carrying the helpless Acting Chief. Olive followed, helping Mahmed to assist Mrs. Shallop, who was uttering unheeded complaints about everybody and everything. To guard against the possibility of any of the party being swept away by the undertow, the halliards had been unrove and were used as a life-line.

It was not an easy passage. The rocks were of coral and irregular in shape, with fairly deep fissures and sharp, jagged crags. Over these ledges the breakers surged, throwing clouds of spray twenty feet or more into the air.

Sounding with the boathook Peter proceeded warily. At frequent intervals he was waist-deep in water. Impeded by the drag of the life-line, half suffocated by the salt-laden spray, and constantly slipping on the kelp-covered rocks, he held on his way, wondering how the others fared, until he gained the dry sand.

The lascars had risen nobly to the occasion. Their solicitude towards their disabled officer was so great that Preston felt very little discomfort. Uncomplainingly they had endured torments from the sharp rocks, that had cut their light footwear almost to ribbons.

Olive Baird had made light of her part of the business, although both she and Mahmed had their work cut out to half drag, half carry the portly figure entrusted to their care. Mrs. Shallop seemed utterly indifferent to the danger and inconvenience of the passage ashore. Her chief anxiety, expressed in peevish accents, was regarding the loss of her "valuable" diamond, which might either be in the boat or else washed through the gaping seams into the trackless waste of sand.

With feelings of thankfulness Peter marshalled his flock under the lee of the cliffs. A hasty examination by means of the lantern resulted in the discovery that the beach was well above high-water mark, so that there was no necessity to undertake the hazardous task of scaling the cliffs in the darkness.

"Where are we, do you think, Peter?" asked Olive. She had dropped the "Mister" quite naturally, since Mostyn had declared his intention of seeing her home.

"Somewhere in Madagascar," replied Peter. "Where, exactly, I have no idea. We'll probably find out from the first natives we come across."

"Are they savages?"

"Hardly. They used to be half civilized only a few years ago, I believe," replied Peter. "Thanks to the beneficent efforts of the French Government, when Madagascar became a dependency of France, they are now orderly and well conducted. Excuse me, Miss Baird, but there are one or two things I have to see to."

Calling to the two lascars, and bidding Mahmed stay with the rest of the party, Peter took the lantern and walked to the water's edge. The tide was fast receding, and most of the ledge was above the water.

Satisfied on this score Mostyn made his way back to the boat, the lascars following. Apparently the stranding had occurred at the top of high water, and the wrecked craft was now perched upon a jagged ledge of coral. She had not altered her position, except for lying well over on her port bilge keel.

In a few minutes the boat was stripped of every piece of movable gear. Twice the salvage party returned to the boat, until nothing was left but the bare hull.

Work for the night was not yet over. By the aid of the masts, sails, and spars, four tents were rigged up under the lee of the cliffs, and a fire was made with the dry kelp and driftwood, augmented by a few detached planks from the boat. A double ration of biscuit and water was served all round, followed by cigarettes for the men and Turkish delight for Mrs. Shallop and Olive. The last commodity came entirely from the latter's share, since the naval officer's daughter had already eaten hers. Yet without the faintest compunction, and looking upon Olive's generosity as a right, the worthless woman had no hesitation in asking her former paid companion for more.

"I'll buy some at the first shop we see," she added, as if Africa's largest island was a hot-bed of up-to-date confectionery stores.

To this the girl made no reply. In fact, she had hardly heeded the fatuous remark. Gazing into the comforting glow of the fire, she was deep in thought as to what the future held in store for the handful of survivors from the S.S. *West Barbican*.

CHAPTER XXVIII
The Island

With the first streaks of dawn, Peter, who had been sleeping soundly in the open, with his feet towards the still glowing embers, shook himself like a great mastiff, and stretched his cramped limbs. It had been a strange sensation sleeping on the hard ground after days and weeks on the ocean. Some moments elapsed before he was fully aware of his surroundings.

He looked seawards. The flood-tide was making, and the wavelets were lapping against the edge of the serrated reef. The boat was still aground. Her anchor warp had not tautened, so that it was obvious that she had not shifted her position on the top of the previous high water.

The wind had piped down considerably, but was now blowing softly from the west'ard. During the night the breeze had veered completely round from east to west.

"Just our luck!" thought Peter. "Now we have fetched Madagascar after beating for hours against it, the wind shifts round. It would have saved us hours if it had been in this quarter for the last twenty-four hours. However, here we are, so I mustn't grouse."

None of his companions showed signs of stirring. Silence reigned in the tents. The scent of the morning air was mingled with the pleasing reek of the camp-fire. Farther along the coast a number of seagulls were hovering over some object and screeching, as they warily circled round the coveted piece of flotsam.

From where Peter stood, the landscape was rather limited. Less than a mile to the nor'ard a bluff of about two hundred feet in height served as the boundary of his vision in that direction. Southward the wall of cliffs terminated abruptly at a distance of about a quarter of a mile. Evidently beyond that the coastline receded, unless the light were insufficient to enable the more distant land to be seen.

"May as well stretch my legs," thought Peter. "I'll have a shot at getting to the top of the cliffs and see what's doing. I wonder how far it is to the nearest village?"

He had to walk a hundred yards along the beach before he found a likely means of ascent—a narrow gorge through which a clear stream dashed rapidly. Yet the rivulet never met the sea direct. The water, although of considerable volume, simply soaked into the sand and disappeared.

"We shan't need to go slow with the drinking-water," he said to himself, as he gathered a double handful of the cool, sparkling fluid and held it to his lips. "By Jove, isn't that a treat after water from a boat's keg. Well, here goes."

The ascent was steep but fairly easy. Nevertheless Mostyn was so out of training, from a pedestrian point of view, that his muscles ached and his limbs grew stiff long before he arrived at the top.

At length, breathless and weary, he gained the summit and threw himself at full length upon the grass.

After a while he stood up and looked around. The sun was just rising— and it appeared to rise out of the sea. From where he stood, Peter could see right across the ground from west to east and from north to south; and, save where the tall bluff cut the skyline, sea and sky formed a complete circular horizon.

Peter gave a gasp of astonishment. Instead of finding himself, as he had expected, upon one of the largest islands of the world, he was on a sea-girt piece of land barely three miles in length and two in breadth. In vain he looked for other land. The extent of his view, assuming that the point on which he stood was two hundred feet above the sea-level, was a distance of roughly twenty miles, and, except for the island upon which the boat had stranded, there was nothing in sight but sky and sea.

"So much for Madagascar," ejaculated the Wireless Officer. "I'm a rotten bad navigator. Wonder where this show is, and if it is inhabited."

For the most part the island consisted of a fairly level plateau covered with scrub. The southern part was well wooded with palms, while the course of the little stream was marked by a double line of reeds.

In vain Peter looked for signs of human habitation. Not so much as a solitary column of smoke marked the presence of any inhabitants.

"This is out of the frying-pan into the fire with a vengeance," said the Wireless Officer to himself. "We've plenty of fresh water, it is true, but precious little to eat. And the boat is beyond repair with the limited means at our disposal. Fire, did I say? We can obtain that, so the possibility of having to eat raw or sun-dried fish is removed."

By this time the rest of the temporary sojourners on the island were astir. From his lofty point of vantage Peter could see the three Mohammedans at their devotions at some distance from the tents. Mrs. Shallop was actually out and about, and had deigned to fetch a balerful of water. Miss Baird had thrown fresh driftwood and kelp on the fire, and was apparently undertaking the frying of some of the fish. Propped up on a roll of painted canvas was Preston, slowly and steadily gutting the herrings before grilling them in front of the fire.

"Hello, old man!" exclaimed Peter, when he rejoined the others and had greeted Miss Baird. "Feeling better?"

"Much thanks," replied the Acting Chief. "Soon be O.K., I hope. And what have you been doing, Sparks?"

"Taking my bearings," said Mostyn. "My festive chum, I've made a hash of things. We're on an island."

"Madagascar is an island," remarked Preston. "So why make a song about it?"

"This isn't Madagascar," replied Peter. "It's a small island. A fellow ought to be able to walk right round it in a couple of hours comfortably."

Preston tried to whistle and failed miserably. The attempt was still too painful.

"You seem fond of putting boats ashore on small islands, old man," he remarked. "How about grub? Seen anything in the edible line?"

"A few coco-palms," announced Mostyn. "I didn't investigate. We may strike oil."

"I'd rather strike grub," rejoined the Acting Chief. "Well, there's one blessing—we've cigarettes."

Breakfast consisted of biscuits, fresh water, and fried fish. It was meagre fare, but the hungry castaways relished it. They could have eaten more, but Peter kept an iron hand on the biscuits, and fried fish without biscuits was neither satisfying nor appetizing.

The meal over, Mostyn set all hands—Preston excepted, by reason of his injuries—to work. He meant to keep everybody employed—even Mrs. Shallop. Idleness breeds discontent and discord, and he had no wish to have either.

The first task was to carry the tents and the small kit at their disposal to the high ground beyond the edge of the cliffs. Peter and the lascars managed the spars and canvas between them, while Olive and Mrs. Shallop carried up the lighter gear. Once she made up her mind that she had to work, Mrs. Shallop became quite energetic, finding her way up the cliff-path with tolerable speed in spite of her bulk. By ten in the morning the whole of the stuff brought ashore had been taken to a spot a hundred and fifty feet above the sea-level, and placed in a sheltered hollow within easy distance of the little stream that Peter had discovered.

While the two Lascars were setting up the tents, Peter and Mahmed constructed a stretcher in order to get Preston to the new camp.

The Acting Chief was practically helpless. At first it was thought that his injuries were confined to his head; but after he had been brought ashore his legs were found to have been crushed, and from the knees downwards the limbs were devoid of any sensation of pain, and the muscles incapable of responding to the dictates of his will.

It required twenty minutes of hard yet cautious work to carry Preston to the top of the cliffs, in spite of the fact that the path was fairly easy for an unencumbered person. The difficulty was for the bearers to keep their burden in a horizontal position, and at the same time maintain their footing. For the greater part of the ascent Mahmed was crouching and holding his end of the stretcher within a few inches of the ground, while Peter was supporting his end on his shoulders and cautiously feeling his way, since it was impossible for him to see where he was treading.

At length Preston was brought to the camp and placed in one of the tents, while his bearers, hot and well nigh exhausted, threw themselves at full length in order to rest and regain their breath.

The next step was to salve the boat. This task required all available hands, for the craft was heavily built of elm.

By dint of strenuous exertions the boat was lifted clear of the jagged coral, and dragged along the ledge and up the sandy beach well above high-water mark.

"That will do for the present," decided Mostyn. "She won't hurt there. We'll have to patch her up and resume our voyage as soon as possible."

He spoke sanguinely, but in his mind he realized that the task was practically beyond the small resources at their command. With the exception of a small rusty hatchet, that was discovered under the floor of the after locker, a knife, and a marline-spike, there were no tools available for the extensive repairs necessary to make the boat again seaworthy.

The time for the midday meal came round only too soon. Feeling like a miser compelled to disgorge his treasured hoard, Peter served out more of his carefully husbanded biscuits. These were augmented by coconuts, which Mahmed and the lascars had obtained from some palms growing close to the camp. Up to the present there were no indications of the presence of bread-fruit trees, but, as Olive remarked, there was a good deal of the island to be explored.

"What's the time, Miss Baird, please?" inquired Preston.

The girl consulted her watch.

"Five minutes to twelve, Mr. Preston."

"Thank you," rejoined the Acting Chief, then, after a slight pause, "is your watch fairly accurate?"

Olive shook her head.

"I never possessed a fairly accurate watch," she replied. "Mine gains about a minute a day, and every time I wind it I put it back a minute. It was set by ship's time on the day the *West Barbican* sank."

"Why so anxious to know the time, old man?" inquired Mostyn. "You haven't to go on watch."

"Never you mind, old son," rejoined the Acting Chief. "In due course I'll enlighten your mind on the subject, but until then—nothin' doin'."

For the next ten minutes conversation drifted into other channels. Peter had almost forgotten about the mysterious inquiries of Mr. Preston, when the latter inquired abruptly:

"What do you think is our position, Sparks?"

"About fifty miles west of Madagascar," replied Peter.

The Acting Chief shook his head.

"Wrong, my festive. Absolutely out of it," he stated with conviction. "Say a hundred and fifty miles to the south'ard of Cape St. Mary—that's the southern-most point of Madagascar—and you won't be far out."

"But, why——?" began the astonished Wireless Officer.

"Hold on," continued Preston. "It's now mid-summer in the Southern Hemisphere. Consequently the sun must be overhead, or nearly so, on the Tropic of Capricorn. Here, at midday, it's roughly five degrees north of our zenith. That means we're well south of the island you were making for."

"But how's that?" demanded Mostyn. "I steered due east, and when the wind headed us I tacked for equal periods."

"Maybe you did," rejoined the Acting Chief drily. "You don't know the deviation of the boat's compass. Neither do I, for that matter. It might be points out on an easterly course. Again, there's a strong current setting southward through the Mozambique Channel. Another and by no means inconsiderable factor is that almost every boat when close-hauled sails faster on one tack than the other. The net result is that, unconsciously, you were faced well to the south-east instead of making due east. However, here we are, and we must make the best of it. Everything considered, old man, you haven't done so badly."

By dusk everything was in order so far as their limited resources permitted, even to the extent of building a light breastwork on the windward side of the camp to protect the tents from storms from seaward. The strenuous labours had kept the castaways' minds so fully occupied that they had had no time to think about their difficulties.

Tired in body, yet cheerful in mind, they slept the sleep that only the healthy can enjoy.

CHAPTER XXIX
Repairs and Renewals

At sunrise on the following morning Peter scaled the highest point of the island, hoping that in the clear air his range of vision would be increased sufficiently to make out land.

He was disappointed. Nothing of the nature of land was in view. The horizon, clear and well defined, surrounded him in an unbroken circle.

He was considerably troubled in his mind over the situation. Desert islands were all very well in their way, provided there was a chance of getting away from them. Evidently this island was well out of the regular steamer track, while sailing vessels, running between The Cape and India and the Federated Malay States, would pass well to the eastward in order to take full advantage of the monsoons.

The boat was practically useless as a means of leaving the island. Had there been a supply of nails in the locker, Peter would not have hesitated to fasten a sheet of painted canvas over the holes in the garboards, and then risked a dash for Madagascar. But without suitable material that was out of the question.

Naturally of an inventive turn of mind, Peter thought out half a dozen plans to make the boat seaworthy; but, as fast as he worked out a solution of the difficulty, objections apparently insurmountable caused him to reject the scheme and start afresh on another tack.

His previous error in navigation rather damped his enthusiasm, but with Preston on the road to recovery he was no longer dependent on himself. The Acting Chief had had years of experience of the Indian Ocean, and, knowing the set of the chief currents and the direction of the prevailing winds, would be of material assistance in navigating the boat—provided she could be made seaworthy.

Still pondering, Mostyn descended from the bluff and walked towards the camp. A more urgent problem demanded his attention: that of catering for the needs of his companions and himself.

The biscuits would not last out much longer, coconuts were unsatisfying fare, and apt to have injurious effect if used as a staple form of food. Whether the island possessed other resources, either animal or vegetable, had yet to be seen. Preliminary investigations had drawn blank in that direction.

Returning to camp, Mostyn found the others busily engaged in getting breakfast. Mahmed had found some oysters, many of them a foot in diameter, while the lascars had surprised and killed a small turtle.

It was rather a curious fact that Mrs. Shallop, childishly ignorant on most matters, was an authority on cooking. She just "took on" the turtle as a matter of course, and by the time Peter returned the choicest parts of the animal were stewing over a wood fire. In the absence of a suitable pot, for the baler was far too small, the self-constituted cook had employed the shell of the turtle as a receptacle for the stew. The oysters were eaten raw, flavoured with the vinegary milk of a young coco-nut.

But the success of the meal was the result of Mahmed's investigations. He had wandered towards the main coco-nut grove on the southern point of the island and had discovered a number of "jack-fruits", a species of bread-fruit. These had been sliced and roasted, forming a good substitute for bread. The lascars, however, disdained the fruit, and were content with the seeds, which they bruised and cooked in coconut shells.

For the moment the grim spectre of starvation had been driven away.

"I've been thinking, Peter," remarked Olive during the rest-interval. "Couldn't we make a canvas boat? We have plenty of sail-cloth, and we could use timbers and planking out of the damaged boat."

"Might," admitted Mostyn. "It would take some doing, and after all it would be a frail craft to carry seven people. We might try it."

He thought over the matter, and the more he did so the greater became the difficulties. Even in calm water a canvas boat, unless properly constructed of suitable materials, is a sorry craft. In the high-crested waves of the Indian Ocean she would not stand a dog's chance.

Yet Olive's suggestion was not without good result. Based upon the idea, Peter's thoughts returned to the damaged boat. Could that not be patched with canvas and strengthened by woodwork so that it would be once more seaworthy?

"By Jove, Olive!" he exclaimed. "I believe you've put me on the right tack. Come on down to the boat. We'll take the lascars with us and see what's to be done. The sooner we get away from this place the better."

Olive did not agree with the latter remark, although she made no audible comment. She was rather enjoying the novelty of the situation. Peter, on the other hand, had got over the glamour of desert islands. An exciting time upon a coral island in the North Pacific had cured him of that. It wasn't to be regretted from a retrospective point of view, but he did not hanker after a repetition.

By the aid of a tackle composed of the halliards and main-sheet blocks the boat was canted over and finally lowered keel uppermost. The full extent of the damage was then apparent. There was a jagged hole about nine inches in diameter through the garboard strake and the strake next to it on the port side about five feet from the stem. On the starboard hand was a smaller hole close to the bilge keel, while there was a slight fracture on the same side eighteen inches from the stern-post.

"Rather a lash-up, what?" exclaimed Peter, as he noted the damage. "Guess we'll be able to tackle that."

He first directed one of the lascars to trim the jagged holes with the axe. The next step was to smooth down the planking adjacent to the gaps by means of canvas and wet sand. This done, the boat was lifted on to her side and the bottom boards removed. A corner of the axe was then employed to remove the brass screws from the stern-sheet benches, while the gratings were sacrificed for the sake of the brass brads that secured them.

This task occupied the whole morning.

After lunch, work was resumed. Strips of painted canvas, smeared with a sticky substance smelling of turpentine, were laid over the holes and tacked down with the brads. Over this canvas the dismembered bottom-boards were firmly screwed. In less than an hour and a half this part of the work was completed.

The boat was then turned over on her keel, and the holes levelled flush with the inside planking by means of clay found in the bed of the little stream. Over this additional canvas was tacked and pressed into position by strips of wood from the bottom boards, struts being fixed between them and the under side of the thwarts to counteract the pressure of the water.

Well before sunset the task of making the boat water-tight was completed, and Peter surveyed the result with intense satisfaction.

"To-morrow," he declared to Olive, who had been working as steadily as anyone, "to-morrow we'll test her. I don't think she ought to leak very much."

"Aren't we going to explore the island, Peter?" asked the girl wistfully.

Mostyn capitulated.

"Yes, certainly, if you wish," he replied. "We can do that easily in a few hours. I don't suppose you'll find it particularly interesting. You see, the weather looks as if it will be fine for some days, and I naturally want to take advantage of it. What do you say to a jaunt before breakfast? We could take something to eat with us, of course. That will leave the forenoon clear for testing the boat."

This suggestion was acted upon, and soon after dawn on the following day Peter and Olive set out on their tour of exploration.

It was a very enjoyable walk for both: to Mostyn because of the companionship of a jolly, unaffected girl; to Olive, because of the novelty of it all. But there was nothing of an adventure about it. The island was devoid of anything of a romantic nature. There were no caves, no traces of former inhabitants. It would have taken a high-flown imagination to weave a thrilling story round that isolated chunk of earth rising out of the Indian Ocean.

They saw no signs of animal life, beyond a few turtles basking on the coral sands, and an occasional lizard scooting for shelter under the trees. There was not a bird to be seen or heard.

Nor did the vegetation give much variety, although Olive discovered a grove of orange trees on the northern extremity of the island. To her disappointment the fruit was intensely bitter and quite unfit to eat.

They returned in time for breakfast, and were greeted warmly by Preston. Mrs. Shallop eyed them with marked disapproval. Although she refrained from making any remark, there was a specially sour look upon her face. Perhaps she regretted having given her companion her dismissal, since by so doing she no longer had control over the girl's freedom.

Directly the meal was over, Peter took one of the lascars down to the beach. It was a perfect day for testing the boat, as the water was as smooth as a millpond, and the tide being full there was little difficulty in launching the repaired craft.

To Mostyn's delight and satisfaction the boat answered admirably. The canvas stood well, and beyond a few drops of water leaking through the seams owing to the action of the sun's rays, the boat was practically watertight.

Quickly the good news was conveyed to the others at the camp, and preparations were begun for the voyage.

Mrs. Shallop had baked quite a quantity of jack-fruit, and had prepared about thirty pounds of turtle-flesh, treating it with brine in order to preserve it for future use. The water-beaker was filled at the stream, and additional water carried in the shells of fully-matured coco-nuts. By two o'clock in the afternoon, just as the north-east breeze sprang up, the camp was struck and the gear stowed away on board the boat.

"Now, old man," said Peter to the Acting Chief; "no mistake this time. You set the course and I'll see that it's kept."

"Right-o!" agreed Preston.

CHAPTER XXX
The Voyage is Resumed

The boat lay riding to her kedge at less than twenty yards from shore. She was in not more than two feet of water. Peter would not risk bringing the boat closer inshore, lest, with her full complement, she would grate over the coral and so injure herself.

Mahmed was first on board, his duty being to assist the two lascars to hoist Preston over the gunwale. This operation was successfully performed without even a groan or a gasp from the injured man, and the lascars returned to carry the portly Mrs. Shallop through the water.

They had a difficult task this time, for the lady confessed to twelve stone, and probably tipped the scale at fifteen. Nevertheless the lascars tackled the job with such a will that their energy was more than sufficient.

Mrs. Shallop began to rock. The oscillations continued until in desperation she clutched at the head of one of her bearers. At the same moment his feet struck a particularly sharp patch of rock. Never "strong on his pins", and additionally handicapped by an unequal share of his fifteen-stone burden, the Indian found himself falling. The prospect of being sandwiched between the sharp coral and the portly mem-sahib was too much for his self-control. With a vigorous and despairing effort he threw himself clear. The other lascar, unable to maintain his charge, let Mrs. Shallop go with a run.

For some seconds she floundered in eighteen inches of tepid water, her horrified features mercifully obscured from the onlookers by a miniature waterspout. Before Mostyn could go to her assistance she regained her feet. For a very brief interval there was absolute silence. Even the lapping of the wavelets upon the shore seemed to have ceased.

Then the storm broke. Mrs. Shallop's pent-up loquacity let itself loose, after being kept under control for nearly forty-eight hours. She stormed at the lascars until they took to their heels, but fortunately they were ignorant of what she did say. Then she directed her battery upon Peter, although he was quite at a loss to know why he should be marked down in this fashion;

while for vehemence her expressions—to quote the immortal Pepys—"outvied the daughters of Billingsgate".

Mostyn suffered the storm in silence. Most people in their passions "give themselves away", and in this instance Mrs. Shallop's outburst simply confirmed Peter's doubts as to the lady's claims to be a naval captain's daughter.

But when Mrs. Shallop included Olive in her revilings Peter's square jaw tightened.

"Enough of this!" he exclaimed sternly. "On board—at once!"

Mrs. Shallop hesitated, trying, perhaps, to find a flaw in the armour of her youthful antagonist. For his part Peter kept his eyes fixed steadily upon the infuriated woman, although he found himself inquiring what he could do to enforce obedience should she prove obdurate.

The tension was broken by Preston's gruff voice. From where he lay in the stern-sheets the Acting Chief could see nothing of what was going on. One ear was covered with bandages, but the other was doubly sharp of hearing. To him a refusal to obey lawful orders was mutiny, whether it came from a dago, "Dutchie", or, as in the present instance, from a blindly angry woman.

"You had one ducking by accident," he shouted. "You'll get another by design—in double quick time—if you don't take your place in the boat."

It was high time, Preston thought, that he had a say in the matter. It was a drastic step to threaten a woman with physical punishment, but there were limitations to the patience and forbearance of himself and his companions. A person of the explosive and abusive temperament of Mrs. Shallop in the boat was not only an unmitigated nuisance but a positive danger. Shorthanded as they were, they could not afford to run the additional risk of being hampered by an irresponsible passenger should they get in a tight squeeze, when the safety of all concerned depended upon coolness, quickness, and unhampered action.

The prospect of another sousing quelled the termagant's spirit. Meekly she waded to the boat and scrambled unassisted over the gunwale.

"Now, Olive!" exclaimed Peter. "To avoid a repetition of part of the performance——"

He lifted the girl in his arms and carried her through the water.

By this time the lascars had returned, and the boat's complement was now complete. The kedge was broken out and stowed, and under oars the repaired craft headed for the open sea, where the dancing ripples betokened the presence of a breeze—and a fair wind at that.

Peter was at the helm, with one hand grasping the tiller and the other shading his eyes from the dazzling sunlight. The two lascars rowed, while Mahmed, armed with the lead-line, took frequent soundings until the boat had drawn clear of the outlying reefs.

"Way 'nough!" ordered Mostyn. "Hoist sail!"

While the Indians were engaged in this operation the Wireless Officer, handing Olive the tiller, made a hasty yet comprehensive survey of the bilges. Except for a slight leaking 'twixt wind and water, the boat seemed absolutely tight. The canvas patches, reinforced as they were with woodwork, were standing the strain splendidly and gave not the slightest indication of leaking. Whether they would withstand the "working" of the boat in a seaway was still a matter that had to be proved.

"What's the course, old man?" asked Peter.

"Keep her at nor'-by-east," replied Preston. "Another thirty-six hours ought to work the oracle."

"It's nearly a dead run," reported Mostyn, after he had steadied the boat on her course.

"So much the better, s'long as you don't gybe her," rejoined the Acting Chief. "Not so much chance of making leeway."

Peter saw the force of this contention, but that did not alter the fact that of all forms of sailing "running" was what he least liked. It soon became apparent that there were others who were of a similar opinion, for, as the boat rolled heavily before the hot, sultry wind, Mrs. Shallop and the lascars were quickly *hors de combat*, showing no enthusiasm when the first meal on board for that day was served out.

Even Olive Baird, used as she was to sailing, felt the motion of the boat uncomfortable. The light breeze was scarcely perceptible, although it was making the sail draw well. Not only was the sun pouring down with considerable strength, but the sea was reflecting hot rays of dazzling light.

Already the island astern was a mere pin-prick on the horizon. Ahead and on either beam was the now monotonous expanse of sea and sky.

Late in the afternoon a shoal of flying fish came athwart the boat's course. Evidently they were being pursued, for they flew blindly, several of them bringing up against the sail and dropping stunned upon the thwarts.

"Dolphins in pursuit, I think," explained Peter, in answer to Olive's question. "I don't know about that, though," he added after a pause. "Look at that."

He pointed astern. Twenty yards away was the triangular dorsal fin of a shark.

"The brute," ejaculated Olive, with a slight shudder. "I hope he goes off soon."

But the girl's wish was not to be fulfilled. If the shark had been chasing the flying fish he no longer did so. Perhaps he scented promising and more satisfying fare, for without any apparent effort he began to follow the boat, rarely increasing or decreasing the distance.

"Hang the shark," exclaimed Peter. "Here, Olive, is a chance to show what a good shot you are."

He handed the girl his automatic. Without hesitation Olive took the somewhat complicated weapon. Peter noted, with a certain degree of satisfaction, that she handled it fearlessly, and at the same time with proper caution. He had no cause to duck his head because of the muzzle pointing in his direction.

"Don't forget to release the safety-catch," he said.

"I've done so already," rejoined Olive, pulling back the mechanism that performed the double action of cocking the pistol and inserting a cartridge into the breech.

It was not an easy target, even at twenty yards. Not only was the boat yawing, but the dorsal fin of the shark was constantly on the move.

The pistol cracked. Mostyn, intent upon preventing the boat from gybing, had no opportunity of seeing the result of the shot. The girl, replacing the safety catch, handed the weapon back to its owner.

"Missed it, I'm afraid," she said. "But there's one good thing—the shark's disappeared."

"Scared stiff, if not hit," rejoined Peter. "Do you mind hanging on to the tiller, while I clean out the barrel?"

The day wore on. At six o'clock Peter roused one of the lascars, and told him to take on for a couple of hours. Already the tent had been rigged amidships, while the jib—useless, or nearly so, while running—had been employed as a sun-screen for Preston.

The sun sank to rest, its slanting rays turning the hitherto blue sea into a pool of liquid, ruddy fire, that gave place to a spangled carpet of indigo as the long undulations reflected the starlight. Away in the west the young moon was on the point of setting. It was the sort of sub-tropical evening that made the discomfort of the open boat pale by its soothing influence.

At eight Peter "took over". He had no desire for sleep, and was quite content to keep watch until relieved at dawn by one of the lascars; but he was somewhat surprised to find that Olive was likewise disinclined to turn in.

They watched the crescent moon dip behind the horizon; they saw the stars pale as a slight mist rose from the waters of the Indian Ocean, and the starlight give place to a darkness broken only by the feeble rays of the binnacle lamp.

By this time the wind had dropped to a gentle breeze on the port quarter, and there was no longer any risk of gybing. The erratic movement of the dead run had given way to the steadier "full and bye", with sufficient "kick" in the helm to make steering a pleasure rather than a monotonous routine.

Suddenly the boat quivered and heeled over to starboard. The shock was sufficient to rouse the sleepers.

"Aground!" exclaimed Olive.

Peter put the helm down. The boat responded instantly to the action of the rudder.

"No," he replied. "We've hit something. Wreckage, perhaps."

"It's a fish!" declared the girl, as with a trail of phosphorescence a huge object darted under the keel and disappeared in the darkness. "That shark."

"Or another one," rejoined Peter. "There's one blessing: it isn't a whale. Chup rao!" he called out to the jabbering lascars.

In two or three minutes the awakened members of the boat's crew had relapsed into slumber. Peter swung the boat back on her course, and handed the tiller to the girl.

"I'll have a cigarette, if you don't mind," he said.

"And one for me, old thing, while you are about it," added a bass voice from the stern-sheets.

"By Jove, Preston, I thought you were sound asleep," remarked Peter, as he placed a cigarette to the Acting Chief's lips.

"Keeping an eye on you, old thing," retorted Preston, with brutal candour, then in a lower tone he added.

"Don't say a word to the girl, but I believe we've sprung a leak. Hear that? It's not the water lapping the boat's sides. It's water trickling in fairly fast. Put a lascar on with the baler. That ought to keep it under until we can see what's wrong."

"Right-o," rejoined Mostyn.

He began to make his way for'ard, moving cautiously past the tent in which Mrs. Shallop was breathing stertorously. But before he could get to the nearest of the three Indians a wild shriek rent the air.

For the moment Peter was under the mistaken impression that he had trodden upon the sleeping form of Mrs. Shallop, but his fears on that score were corrected by the lady exclaiming:

"We're sinking. I'm in the water. Let me out! Let me out!"

It was some time before the Wireless Officer could release the woman. She had laced the flap of the improvised tent from the inside, finishing up with a wondrous and intricate knot. In the darkness the task was even more difficult. Peter solved it by wrenching one side of the canvas away from the gunwale, and was rewarded by being capsized by the impact of Mrs. Shallop's ponderous and decidedly moist figure.

Meanwhile Mahmed, acting upon his own initiative, had lighted the lamp. By the uncertain light Peter found that his fears were realized. Water was spurting in through a rent in the canvas patch on the gar-board strake.

A long, pointed object attracted his attention. It was the beak of a large sword-fish. The creature had come into violent contact with the boat, driving the formidable "sword" completely through the temporary planking, two thicknesses of heavy canvas, and the intervening padding of clay. The bone had broken off short, but the worst of the business was that the sudden wrench had split the piece of elm forming the outside of the patch, and through the long narrow orifice thus made, gallons of the Indian Ocean were pouring into the boat.

Desperately Peter strove to wrench the sword clear of the hole. It swayed easily enough, but no amount of force at the Wireless Officer's command enabled him to remove the long, tapering horn.

"Bale away!" he exclaimed to the lascars, who were inertly watching their sahib's efforts to free the swordfish's formidable spike. "Bale, or we'll sink."

"If you can't pull it out, push it back, old son," exclaimed Preston.

Glancing up, Peter found the Acting Chief in a sitting position, supporting himself with one hand grasping the after thwart.

Mostyn acted upon the advice, but he proceeded warily. It was a fairly easy matter to knock out the sword with a metal crutch—it was merely driving out an elongated wedge—but the question arose whether any display of force would prise the temporary planking from its fastenings.

At last to his satisfaction he felt the horny spike giving. After that it moved easily. Peter pushed its point completely clear of the boat, but the next instant the water poured in with redoubled violence, a phosphorescent waterspout rising a good eight or ten inches above the kelson.

Seizing a piece of canvas Peter wedged it into the gaping hole. The inflow was appreciably checked, but in order to withstand the pressure it was necessary for some one to hold the "stopper" in position, until repairs of a more substantial nature could be effected.

Calling to one of the lascars, Peter bade him carry on with the plugging process.

Hot, wellnigh breathless, and spent with his exertions, Peter sat up. He glanced aft. The feeble light from the binnacle showed him that Olive was at the helm, calm and collected. Throughout the anxious five minutes she had kept the boat on her course with the skill of a master-mind—a vivid contrast to the hysterical woman whose incapacity in a tight corner belied her oft-repeated statement as to her naval forbears.

And during that five minutes the breeze had freshened considerably. Already the seas were breaking viciously, their white crests showing ominously in the darkness. Another peril faced the crew. Could the badly strained and leaking boat withstand the onslaught of the threatened storm?

CHAPTER XXXI
Picked up at Sea

"I'll attend to the leak, Peter," volunteered Olive. "That will leave you free to shorten sail."

"Topping!" exclaimed Mostyn. "Keep your foot on that pad of canvas. Don't press too hard or the whole gadget may carry away."

Reefing was a difficult matter, for the boat was driving heavily and the canvas was as stiff as a board. Mostyn dared not risk lowering the sail. The little craft had to carry way to prevent her broaching-to and being swamped. It seemed incredible that in the short space of five or six minutes the hitherto calm sea should have worked up into a cauldron of crested waves and flying spindrift.

In the contest with the elements Mostyn temporized. Putting the helm up slightly and easing off the sheet, he released the pressure on the canvas sufficiently to enable Mahmed and the two lascars to take in a couple of reefs. At the same time the boat was travelling fast but was well under control.

"Let's hope it won't blow any harder," thought Peter. "She won't stand much more wind, and she'd break her back if she had to ride to a sea-anchor."

One of the lascars came aft and reported that the reefing operation was complete. Peter put the helm down to bring the boat back on her course, when, with a report of a six-pounder quick-firing gun, the tightly stretched canvas parted. Cloth after cloth was rent in rapid succession until the severed sail streamed banner-wise before the howling wind.

Somewhat to Mostyn's surprise and satisfaction the boat showed no inclination to broach-to. Possibly the fluttering canvas offered sufficient resistance to the wind to enable her to answer to the helm.

The next task was to set the jib as a trysail. It was almost useless to expect the lascars to do that. Their knowledge of boat-sailing was very elementary, having been gained in handling their native craft, and occasionally the ship's boats under regulation rig and in charge of their British officers.

Ordering Mahmed to take Miss Baird's place at the leaking patch, Peter handed the tiller over to the girl. There was no need to caution her as to what was to be done. She knew perfectly well that safety depended upon her ability to keep the boat's stern end on to the following seas.

Mostyn had no fears on that score. He knew the girl's capability in that direction by this time. Thanking his lucky stars that he was not dependent upon the indifferent seamanship of the lascars, he went for'ard with the jib which Preston had to relinquish as a covering.

In almost total darkness Peter found the head and tack of the sail. Fortunately the split mainsail was still held by the luff ropes, thus enabling him to gather in the fiercely flogging fragments and secure the lower block of the main halliards.

To the latter he bent the head of the jib. It was now a fairly easy matter to hoist the diminutive triangle of canvas and sheet it home.

"She'll do," he exclaimed, as he relieved Olive at the helm.

The girl nodded in reply. She was too breathless to speak. Her brief struggle with the strongly kicking tiller had required all the strength at her command. There was, she discovered, a vast difference between the long tiller of a well-balanced sailing dingy on the sheltered waters of the Hamoaze, and the short "stick" of a heavy ship's boat on the storm-tossed Indian Ocean.

Through the long hours till morning the boat ran before the storm. Never was day more welcome. At dawn the wind piped down and the sea moderated. The boat had made a fair amount of water, not only through the leaking patch, but over the gunwale, and, in order to keep the leak under, one of the lascars had to keep his hand down on the canvas stopper while the other plied the baler. This they had to do turn and turn about throughout the night, and by dawn they were both pretty well done up.

By nine o'clock, when the sun had gathered considerable strength, the wind had practically died away, and the sea had resumed a smooth aspect save for a long, regular swell. Only a few ragged wisps of canvas and the now almost idle and ridiculously inadequate trysail remained as a reminder of the night of peril.

In vain Mostyn looked for signs of land. Nothing was in sight save sea and sky. To make matters worse, the boat, which in that light breeze would have made about three knots under her mainsail, was now barely carrying steerage way. At that rate she might take weeks to fetch land—if she ever did so at all.

Breakfast over—it was a more substantial meal than their previous ones in the boat—Mostyn set the lascars to work to rig up jury canvas. The damaged mizzen-sail, that had served as a tent, was pressed into service, together with the tarpaulin. These were "bonnetted" together, bent to the gaff, and sent aloft as a square sail, with the result that the boat's speed increased perceptibly. Yet there was still a great difference between her normal rate and that under the jury canvas.

Smoking a cigarette after the meal, Peter let his thoughts run riot. He wondered what his parents were doing; whether they had had by this time any report of the *West Barbican*. If so, were they mourning him as dead?

"Rather rough luck on them," soliloquized the youthful optimist; "but won't they be surprised when I roll up again?"

Then his thoughts went to the Brocklington steel contract. He wondered whether the Kilba Protectorate officials had sent to Bulonga for the consignment. It seemed to him rather an idiotic thing to do, to have the stuff dumped down in that out-of-the-way hole, when the *West Barbican* might, with equal facility, have delivered it at Pangawani. Perhaps, after all, it was for the best. The stuff might have gone down in the ship, in which case Captain Mostyn would be a ruined man.

The mysterious loss of the *West Barbican* had been a source of frequent perplexity to Peter. He was thinking about it now, trying to put forward a satisfactory theory as to the cause of the explosion. As far as he was aware there were no explosives on board, a consignment of gelignite, for use on the Rand, having been landed at Durban.

His reveries were interrupted by one of the lascars shouting: "Sail on port bow, sahib!"

Peter sat up. The foot of the improvised square sail intercepted the view for'ard. It was not until he made his way to the bows and stood upon the mast thwart that he saw the craft which the lascar had indicated.

She was still a long way off, only her canvas and the upper portions of her hull showing above the sky line. At that distance it was impossible, without the aid of a telescope or binoculars (neither of which was on the boat), to distinguish her rig or in which direction she was heading. As she was a sailing craft, and, taking for granted that she carried the same wind as the boat, the chances were that she would soon disappear from sight.

Nevertheless Mostyn meant to leave nothing undone that might attract the stranger's attention. Rockets were fired in the hope that the loud detonation might be audible at that distance. The light they gave out would be unseen in the terrific glare of the sun.

At Preston's suggestion strips of canvas were soaked in lamp oil and set alight at the end of the boat-hook. These flares gave out a dense smoke that rose to an immense height in the now still and sultry air.

For the best part of half an hour these signals were repeated at frequent intervals. Then, to everyone's disappointment, the strange sail faded from view.

"It's not to be wondered at," remarked Preston. "You know what a look-out at sea is like; and, in any case, they don't keep a fellow on watch to see what's coming up astern."

"They ought to," declared Olive.

The Acting Chief was sitting up, his back supported by some spare oilskins folded over the after thwart.

At the girl's retort he winked solemnly with the eye that was not covered with bandages.

"Do we?" he asked. "Look astern now."

To the surprise of everyone else in the boat a large sailing craft was bowling along dead in their wake. She was now a little less than a mile away, and had evidently been attracted by the signals made to the craft that had so recently been sighted in vain.

"A rum sort of packet, by Jove!" exclaimed Peter.

"A dhow, my sweet youth," explained Preston. "'Tisn't often you find 'em so far south, but you'll see shoals of them up along the coast from Mozambique and Zanzibar right up to the Red Sea and Persian Gulf. Clumsy-looking hookers, but they can shift."

It was Mostyn's first sight of an Arab dhow. He had seen plenty of Chinese junks in Shanghai whilst he was on the Pacific trade. This craft reminded him of them, only its rig was more in accord with Western ideas. End-on it was impossible to see that the masts raked at different angles, but the well-drawing lateen sails and the "bone in her teeth" indicated that she was a swift craft ably managed. Even in the light air she was moving at about six knots.

The Wireless Officer leant forward and whispered in Preston's ear.

"S'pose she's all jonnick, old man?" he asked.

"Sure," replied the Acting Chief. "The slave-dhow and the gun-runner are as dead as the dodo in these parts. Probably she's a trader from Reunion, blown out of her course by the late hurricane. Nothing to worry about, old son."

"Right-o!" rejoined Mostyn, and ordered the lascars to lower the sail and to stand by with the painter.

By this time the dhow, which was coming up "hand over fist", was about a cable's length astern. From the boat it was impossible to see the helmsman of the overtaking craft, owing to the foot of the lateen sail, but in her low bows could be discovered three Arabs intently looking in the direction of the now motionless little craft.

Presently a high-pitched voice called out an order. The hitherto listless Arabs for'ard sprang into activity. With a smartness that would have evoked admiration from the most exacting seaman, the lateen yards were lowered and squared fore and aft, while the dhow, still carrying way, ranged alongside the *West Barbican's* boat.

"Any port in a storm," thought Peter, as the lascar for'ard threw the painter into the hands of one of the Arab crew. "I wonder what we're in for now?"

CHAPTER XXXII
The Dhow

Mostyn was the first to board the succouring craft. Somewhat dubious as to the nature of his reception, he swarmed up the low side and gained the deck.

His arrival elicited no demonstrations, either of friendliness or hostility, from the white-robed Arabs. They simply looked at him without visible signs of curiosity; without even the formal salaam.

There were five of the dhow's crew. Four, who had been attending to the lowering of the sails, were standing amidships; the fifth, a mild-looking, bearded man of more than average height, was at the long, curved tiller. Save for his swarthy skin he might have passed for a European, for his features were regular, his nose aquiline, and his lips red and without the fullness of the typical African. He wore the white "jebbah" and burnous, the only dash of colour being his red Morocco slippers. In his white sash could be seen the leather-covered hilt of a long knife.

"English," explained Peter. "Wrecked—want passage."

The Arab shook his head gravely, and motioned to Mostyn to get the rest of the boat's party on board.

"Mahmed!" sang out his master.

"Sahib?"

"You speak Swahili. Tell this man who we are and what we want."

Mahmed came over the side and approached the Arab captain. Apparently the former's attempt to speak Swahili was far from fluent, but the desired result was obtained.

"He for Dar-es-Salaam, Sahib," explained Mahmed "He promise passage one hundred rupees a head."

"He'll get it," replied Peter. "We'll give him one thousand rupees if he puts into Pangawani."

The Arab rejected the amendment. He was willing enough to give them a passage, but he was not going to put into an intermediate port even for the inducement of an addition three hundred rupees.

Preston was the next to board the dhow. He managed it practically unaided, for his lower limbs were regaining strength, and he was able to use his left arm. The Arabs showed considerable interest at his bandaged head, the captain going to the length of inquiring of Mahmed how the injuries were caused.

Mrs. Shallop and Olive followed.

The two lascars completed the transhipment. They brought with them the scanty personal belongings of the party, together with the water-beaker and the rest of the provisions.

"Tell him we are ready to cast off," said Peter.

Mahmed translated. The Arab skipper went to the side and cast envious looks at the boat, for from the deck of the dhow the damaged planking was not visible. With an instinct not confined to dhow-owners he was loth to abandon a craft that Providence had figuratively thrust into his hands; but upon consideration he was compelled to admit that the gift was too unwieldy. Nevertheless, since he was unable to make use of the boat, he was determined not to give others a chance of so doing.

At his order a couple of Arabs, armed with knives and small-headed axes, jumped into the boat. After removing the compass, oars, masts, and remaining sails, and all other loose gear, they cut the gunwale through to the water-line, regaining their own craft as the water poured through the jagged rent. The painter was cut as close to the boat as it was possible for a man to reach from the dhow, and the *West Barbican's* boat, her mission accomplished nobly in spite of difficulties, drifted slowly astern in a water-logged condition. Then, the lateen sails rehoisted, the dhow resumed her course, hauling close to the wind on the starboard tack, her head pointing practically nor'-west-by-north. For the best part of an hour the survivors of the *West Barbican* remained on deck, no attempt being made on the part of the Arabs to offer them accommodation and shelter below. The captain had handed over the helm to one of the crew, and with the other three men was squatting on the deck. There was apparently no social distinction between the Arab skipper and his crew. They were eating *pilau* from a common dish, and talking loudly, as if oblivious of the presence of the "Kafirs" and the three Moslem members of the rescued party.

At length Peter thought it was time to assert himself on behalf of his companions. It was scant comfort to have to grill upon the deck of the dhow, for the sails provided little shelter from the fierce rays of the sun.

Calling to Mahmed to accompany him, Mostyn made for the short ladder giving access to the steeply shelving poop.

Seeing Peter's intention the Arab captain stood up and warned the intruder off, at the same time talking angrily to the Indian interpreter.

"Tell the accursed Kafir not to set foot upon the ladder," was what he said, but translated by Mahmed the message was, "The sahib is kindly asked not to approach while the crew are having a meal."

Which was unfortunate. Out of deference to Arab customs Peter complied with the request. The captain took it for a sign of weakness on the Englishman's part. Had Mahmed translated literally, Mostyn would have been on his guard. It would have been clear that the Arab had not any intention of setting the party ashore at Dar-es-Salaam or at any other port where the British flag was flying, otherwise he would never have dared to insult a man who was quite capable of turning the tables on him on arrival at a place within the sphere of British influence.

Mostyn waited more or less patiently until the *pilau*-eating party had broken up. Then he again approached the Arab skipper, who was now standing at the head of the poop ladder.

The Arab avoided a reply to the direct request for shelter by demanding immediate payment of the seven hundred rupees.

"Tell him," said Peter, "that the money will be paid directly we arrive at Dar-es-Salaam."

A faint smile fluttered over the Arab's olivine features.

"Has the Kafir the money with him?" he asked.

"That has nothing to do with the bargain," replied Peter, through his interpreter. "He will be paid promptly and in full when he has carried out his part of the deal, but for that sum we must have suitable accommodation."

For a while the Arab looked decidedly sulky. Then, with another smile, he gave a perfunctory salaam and shouted an order to two of his crew.

The latter promptly disappeared under the poopdeck, where they spent some time shifting gear from one place to another.

When at length they reappeared, the captain led Mostyn to a fairly spacious but low-roofed cabin on the port side of the dhow, and immediately abaft the poop bulkhead.

"That will do for the women," thought Peter. "Now for a place where we can sling our hammocks."

His request through Mahmed for additional accommodation was curtly turned down on the score that it was impossible. Already two of the Arabs had been turned out of their quarters to make room for the Kafirs.

"We won't kick up a shine over that," decided Peter. "Preston and I can have a shelter on deck. We have a right to make use of our own sails. I suppose the women will be safe down here? No lock on the door, but I can show Olive how to jamb it with the blade of an oar. Now there are the lascars to fix up."

That difficulty was quickly settled, the two lascars agreeing to the Arab's suggestion that they should take possession of a small cuddy for'ard, access to which was gained by a small, square hatch just for'ard of the raking foremast. Mahmed, at his own request, was to remain with his master and Preston.

Olive and Mrs. Shallop were duly shown the quarters assigned to them. The latter, for a wonder, raised no objection to the place. Peter could not help thinking that perhaps her overbearing nature had been thoroughly cowed by the rebuff she had met with on re-embarking in the boat.

It was Olive who took exception to the place.

"I think, if you don't mind," she said, "I'll get you to rig me up a shelter on deck. It's rather stuffy down there for two. You have no objection, I hope, Mrs. Shallop?"

"Not in the least," replied the lady loftily. "It's nothing to do with me. You can please yourself."

"Thank you," said the girl promptly.

Peter concurred. Although he was curious to know why Olive should have objection to the cabin—it had been swept out—he refrained from asking why. He could only come to the conclusion that Olive was reluctant to be in her late employer's company more than was actually necessary.

"It was stuffy down there," declared the girl. "No scuttle—I'd much prefer a canvas screen on deck."

The rest of the day passed uneventfully. About four in the afternoon land was seen broad on the starboard beam. What land it was Peter had no idea. The Arabs were silent upon the subject. Preston could advance no suggestion beyond the theory that it might be Cape St. Mary, on the southernmost extremity of Madagascar.

"If so, old man, we were all out of it," he added. "On the course we were steering we would have missed the whole island. Strange things happen at sea."

At sunset the Arab crew turned their faces towards Mecca and prostrated themselves on the deck. In their acts of devotion they were joined by the lascars.

"Black heathens!" snorted Mrs. Shallop contemptuously, laughing loudly.

It was the act of an uneducated fool. People of that type, both male and female, have done so before to-day, often with serious results to themselves and others.

"For Heaven's sake shut up!" hissed Preston apprehensively. "You may get a knife across your throat for this."

Peter too felt far from comfortable when the Arabs regained their feet. There could not have been the slightest doubt that they had heard the mocking laugh, and had there been trouble the lascars would have held aloof, or even have sided with their co-religionists. But, grave and inscrutable, the crew of the dhow carried on as if the unseemly interruption was beneath their notice.

"I think I'll keep watch to-night after this," said Peter.

"P'raps 'twould be as well," agreed Preston. "That woman is a perfect curse—I'm not much use, but I'll take a trick. If there's any sign of mischief I can give you a shout. Got your automatic handy?"

"Rather."

"Pity you hadn't shown it, off-handed like," continued the Acting Chief. "A little moral persuasion of that description goes a long way with these gentry. I remember once getting into a jolly tight corner at Port Said. It was my own fault to a great extent, but I was only an irresponsible 'prentice in those days. I shifted a dozen low-down Arabs with the stem of a pipe. They thought it was a six-shooter. It's as likely as not that our friend the skipper has spotted that bulge in your hip pocket."

"And taken it for a purse with a thousand rupees in it," added Peter. "Yes, I think I'll have to keep my weather eye lifting."

Preston and the Wireless Officer had rigged up a canvas shelter amidships, spreading the covering ridge-wise on a gantline stretched between the mainmast and the for'ard end of the hatch. The hatch was a large one, measuring roughly thirty feet by ten, and was covered with canvas held down by bamboo battens. This, with the tent, took up the greater part of the deck space amidships.

Farther aft, but on the centre line, a tent made from the boat's mizzen sail had been set up for Olive's use. Provided the weather remained fairly quiet it formed quite a sheltered and comfortable retreat.

The Arab captain had raised no objections to the execution of this plan, although it had been carried out without his sanction. Peter and Mahmed had set up the shelters without any hesitation. The former was, indeed, prepared to assert his right to do so in consideration of the fact that he had not pressed his claim for more accommodation under the poop-deck.

It was late before Mostyn turned in. For quite an hour he had stood on deck with Olive, watching the moon sinking lower and lower in the heavens until it dipped beneath the horizon.

Peter gave no hint to the girl of his misgivings, nor did Olive refer to her reasons for refusing to share the cabin with Mrs. Shallop. After all, knowing the lady, he was not surprised at the cultured girl's reluctance to be in her company more than was absolutely necessary.

At about ten o'clock Peter bade the girl good night. Creeping in under the flap of his shelter he found Preston fast asleep on one side of the deck-space and Mahmed, equally somnolent, lying right across the entrance. He stirred as Peter made his way over him, but instantly fell asleep again.

"Fortunately I'm not sleepy," thought Mostyn, as he settled himself upon his share of the rough bedding, which consisted of oilskin coats and a rafia mat.

On deck all was quiet, save for the occasional creaking of the blocks and the ripple of water at the dhow's bows. With the exception of the helmsman the Arab crew had gone below before Peter had retired to his shelter-tent. The lascars had also retired to their assigned quarters for'ard.

The night was calm and sultry. At twelve the solitary watch on deck was relieved; it apparently being the custom on board the dhow for the helmsmen to work three hour-tricks both by day and night.

Peter heard the two men talking for a few minutes in a low tone; then the Arab off duty went below, his slippers pattering softly on the deck.

Another hour passed. Nothing of an unusual nature happened. Mostyn began to wonder whether his precautions had been in vain. He was feeling a bit sleepy by this time, but he had no desire to arouse his injured companion.

He was content to take Preston's word for the deed, but if he were to keep awake he simply must have some fresh air.

With this purpose in view Peter crept cautiously across the sleeping Mahmed, drew aside the flap of the tent, and gained the open air. It was now a fairly bright starlit night. The cool breeze thrummed tunefully through the scanty rigging, gently filling the huge, triangular, lateen sails. The foot of the mainsail was cut so low that from where Mostyn stood, just abaft of the foremast, the shelving poop was hidden from view.

Bareheaded and lightly-clad he grasped one of the weather-shrouds and drunk in great draughts of the ozone-laden air. He realized the relief of being no longer responsible for the safety of his charges, so far as seamanship and navigation were concerned. Day after day, night after night in an open boat had considerably dimmed his ardour for exercising command.

After a while he wanted a cigarette, but remembered that he had left his share in the breast-pocket of his drill tunic.

"Better be turning in again," he soliloquized, with visions of malaria in his mind. "It's rather a risky game hanging about here."

Even as he turned to regain the shelter a shriek rent the air. Less than ten feet from where he stood were a couple of Arabs kneeling beside the collapsed tent. One was holding the canvas down with hands and feet, while the other, knife in hand, was raining furious blows upon the defenceless and sleeping men pinned beneath.

CHAPTER XXXIII
A Fight to a Finish

A mad fury seized upon the Wireless Officer. Without giving a thought to the automatic pistol in his hip-pocket he hurled himself upon the treacherous Arabs.

Strong, agile, and carrying weight, his sudden and unexpected onslaught took the pair as completely by surprise as their murderous attack had taken their victims.

With a crashing blow from his left Peter felled the fellow with the knife, stretching him insensible upon the deck and hurling the glittering steel into the lee scuppers.

So headlong had been Mostyn's rush that its impetus proved his undoing. His foot caught in the folds of the canvas. He tripped across the limp and inert body of one of the occupants of the overturned tent, and with a dull thud he measured his length upon the deck.

He regained his feet quickly, but not before the second Arab had recovered from the shock of the unexpected diversion. The next moment Peter and the Arab were wrestling furiously.

With a mighty heave the Wireless Officer swung his lithe and muscular antagonist from the deck, but the Arab's fingers were gripping Peter's throat in a sinuous and tenacious hold. Swaying, turning in short circles, the two combatants struggled. It was a question of who should be able to hold out longest—the Englishman with his windpipe almost closed or the Arab with his ribs strained almost to bursting-point and his lungs as empty as a deflated tyre.

Once Peter swung the Arab round in the pious hope that he might crash his opponent's head against the mast, but the fellow, although on the point of suffocation, contrived to turn aside. Then with a sudden movement he released his grip on the Englishman's throat, transferring his attention to Mostyn's eyes.

Peter's fairly long hair afforded a secure hold for the Arab's fingers, while his thumb slithered down Mostyn's forehead preparatory to the typically Arab trick of gouging out his opponent's eyes.

"Would you?" spluttered Peter.

Releasing his hold of his foeman's body, he put a rallying effort into a terrific uppercut. The blow was well-timed. The Arab was simply lifted from the deck. His arms outstretched, his fingers still grasping a generous helping of Peter's hair, he described a perfect parabola, Arab Number Two thudded unconscious upon the deck by the side of his previously vanquished compatriot.

Dazed and breathless, Peter strove to recharge his lungs. He was barely conscious of the blood flowing from the raw patches whence his hair had been uprooted. It was his throat that pained terribly. He seemed still to feel the claw-like fingers pressing remorselessly into his windpipe. Every gasp of air rasped his lacerated tongue, which, in his imagination at least, had swollen until it threatened to complete the choking process that his opponent had failed to achieve.

The respite, agonizing though it was, was a short one. A warning cry—whence it came Peter knew not—put him on the alert.

Approaching with swift, cat-like movements were two more Arabs, one of whom was the captain of the dhow. The latter had a knife in his hand, its long blade shimmering in the starlight. The other fellow, although he wore a knife in his sash, relied upon an iron bar as a weapon of offence.

For the first time during the encounter Peter remembered his automatic. The thought gave him confidence for the renewed struggle, but his fingers, trembling with the muscular reaction, fumbled as he drew the pistol from his pocket.

He was a fraction of a second too late. Before he had time to level the weapon the Arab with the bar dealt him a terrific, flail-like blow. Stepping aside and stooping, Peter avoided the swing of the weapon by a hairbreadth, but the automatic was struck from his grasp and flew half a dozen yards along the deck.

The Arab, carried half-round by the impetus of the swing of the bar, finished up by dealing the captain a heavy blow upon the wrist that caused him to drop the knife.

Instantly Peter saw and seized his opportunity. Grasping the Arab sailor round the waist he advanced upon the captain, using the former as a shield and battering-ram.

Retrieving the knife with his left hand, the skipper of the dhow advanced cautiously, to be confronted at every approach by the struggling, helpless form of his compatriot.

TWO TO ONE

It was a strenuous task for Mostyn. Already sorely tried by his previous and successful combat, he realized that the unequal struggle could not last much longer. The weighty and frantically kicking Arab was surely wearing out his last remaining strength, while the comparatively uninjured captain was awaiting his opportunity of rushing in and knifing the exhausted Englishman.

Peter had "seen red", now he was beginning to "see white", for a mist swam in front of his eyes. He felt his knees giving way under him. He was no longer able to hold his human buckler clear of the deck, and the Arab's bare heels were beating an erratic tattoo on the planks.

Seizing his chance, the Arab captain sprang. The steel glittered in the starlight. Peter could see that. He braced himself to receive the stroke, when a dazzling reddish flash stabbed the air, followed almost simultaneously by a loud report.

As far as Peter was concerned the fight was finished. He lay unconscious on the deck, sandwiched between his living buckler and the body of the treacherous captain of the dhow.

CHAPTER XXXIV
Olive deals with the Situation

A violent slatting of canvas was the first comprehensible sound that greeted Peter's ears as he began to recover his senses.

He opened his eyes and stared perplexedly at a light. It came from a familiar object—the boat's lamp. He could not understand why the sails were shaking, unless for some reason the boat had been allowed to run up into the wind, which was great carelessness on some one's part, he reflected.

Yet, somehow, he wasn't in the *West Barbican's* boat, but on the deck of something far more spacious.

He tried to sit up. The movement was a failure, resulting in a throbbing pain in the region of "Adam's apple". Remaining quiet for a few minutes he racked his bewildered brains to find a solution to the mystery.

He was lying on his left side, his head supported on a folded coat. His forehead was bound round with a wet cloth. Why he knew not. It wasn't his head but his neck that was giving him pain.

And what was the boat's lantern doing there?

Then he became aware of a hand touching him lightly on the forehead. He recoiled at the touch, and, turning his head, saw Olive kneeling on the deck beside him.

"Hello!" he exclaimed feebly. "Where am I?"

"Still on the dhow," replied the girl. "You—we—are all right now."

"Are we?" rejoined Peter, still mystified. "Why is she run up into the wind? Can you give me a drink of water?"

Mostyn drank with difficulty. The liquid was refreshing to his parched tongue and lips, although it was a painful task to swallow. Then he looked at the girl again.

Her face was deathly pale, even in the yellow glare of the lantern. She was bareheaded, her hair, loosely plaited, falling over her shoulders. There were dark patches on the hem of her badly worn skirt.

Then in a flash Mostyn remembered everything up to the time when he had lost consciousness—the treacherous attack upon his sleeping companions, his double fight against the four Arabs. Where were they now?

He staggered to his feet, and would have fallen promptly had not Olive held him up. Carefully she piloted him to the coaming of the hatch.

Although Peter's bodily strength was slow of recovery his brain was rapidly regaining its normal functions. Seated on the hatch, with the cool breeze fanning his face, he was able to take stock of his surroundings.

The dhow was not under control. Her lateen foresail was aback. The masterless tiller was swaying to and fro as the vessel gathered stern way.

Close to the mainmast were the disordered folds of the tent, on which lay the motionless forms of Preston and Mahmed. Reclining against the short poop-ladder was Mrs. Shallop, her brawny arms bared to the elbow, and her black hair grotesquely awry. Peter could have sworn that she was wearing a wig.

Neither the two lascars nor the Arabs were to be seen, but the disordered, blood-stained deck bore traces of the desperate fight, while lying close to the fife-rail of the foremast was Mostyn's automatic.

"Are they dead?" inquired the Wireless Officer, pointing to the bodies of the Acting Chief and Mahmed. Somehow he could not bring himself to mention them by name.

"Mr. Preston's got a knife-thrust in the shoulder," replied Olive. "Mahmed has half a dozen wounds, but he's still living. We dressed their injuries as well as we could—Mrs. Shallop and I."

"And where are the lascars?"

"Locked in for'ard," announced the girl. "We thought we would let them stop there a bit until we sorted things out. The Arabs? Mrs. Shallop attended to them. I helped a bit. She wanted to throw them overboard. We lowered them into the after hold—all five."

Peter swallowed another draught of water. He suspected, not without reason, that he presented a pretty sight in the starlight. His shirt had been split across both shoulders, his right knee showed through a long rent in his trousers. His hair was matted with dried blood; his face was scratched and his neck swollen and purple-coloured. In addition, he was bespattered with the blood of at least one of his vanquished antagonists.

"We may as well release the lascars," he said "It's about time we got the dhow under control."

Together Olive and Peter went for'ard and cut the lashings that secured the forepeak hatch. It was quite a considerable time before the lascars summoned up courage to appear, not knowing what had happened, although they had heard the struggle and guessed what was taking place. Fortunately they guessed wrongly. They were not in the power of the ferocious Arabs, and their relief was plain when they realized that Mostyn Sahib was still in command.

Fortunately both men were acquainted with the management of a dhow. The foresail was filled and the helm put up, and once more the unwieldy craft was set upon her course.

There was little or nothing to be done for Preston and Mahmed. The former had recovered consciousness, having sustained a clean cut in the shoulder. It was Peter's servant who had borne the brunt of the initial attack, the Arabs, ignorant of his presence in the tent, having been under the impression that they were knifing his master.

Already Olive and Mrs. Shallop had washed their wounds and bandaged them with the cleanest linen obtainable, which happened to be the burnous of the Arab captain.

"Now you must sleep, Peter," said the girl authoritatively, after Mostyn had done his best for the dhow and her new crew. "You'll be fit for nothing to-morrow if you don't. No, I won't tell you anything more now. We'll be quite all right."

Mostyn obeyed the mandate. Apart from being utterly fatigued he rather liked being ordered about by the self-possessed and capable girl. In default of suitable bedding and covering, for the well-tried sail had been hacked almost to shreds, he stretched himself on a clear space of deck and was soon sleeping the sleep of exhaustion.

When Peter awoke it was broad daylight. Olive was not to be seen, but Mrs. Shallop had evidently been asserting herself—this time to good purpose; for, strange to relate, she was at the helm, while the lascars were engaged upon the finishing touches of "squaring up" the deck.

All traces of the encounter had been removed, and the planks had been scrubbed and washed down. Preston and Mahmed had been carried into one of the cabins under the poop-deck, where already the Arabs' former quarters had been "swept and garnished".

Seeing Peter stir, Mrs. Shallop threw him a curt greeting, with the additional advice that if he went aft he would find something to eat.

Mostyn took the hint. He was feeling peckish. As he stooped to clear the break of the poop he heard the woman shouting to the lascars to "get a move on, as I don't want to hang on here no longer than I can help"—a contradiction of terms which, however, had the desired effect upon those for whom it was intended.

In the aft cabin Peter found Olive presiding over a charcoal brazier and a brass coffee-pot, from which fragrant and almost forgotten odours were issuing. The dhow's larder had been raided, with the additional discovery of dates, dried goat's-flesh, bread, and several commodities of doubtful origin.

Peter enjoyed the meal immensely in spite of his inflamed gullet. Then, over a cigarette, he heard Olive's account of her part in the desperate fight.

It appeared that the Arabs failed through a lack of concentration in their initial attack. Instead of four of them dealing with Peter and Preston (one of the crew had to be at the helm) two crept towards the tent in which the Acting Chief and Mahmed were sleeping while a third secured the hatch over the lascars, and the fourth directed his attention upon the cabin in which Mrs. Shallop had taken up her abode.

Awakened by the uproar, Olive slipped out of her shelter, and hid in the angle made by the rise of the poop and the adjoining bulwark. The place was not only in shadow; it was hidden from the view of the Arab at the helm.

Horror-stricken, the girl watched the drama until she saw that Peter had thrown himself upon the would-be assassins. Up to that moment she had thought that he was struggling under the folds of the overthrown tent.

Then horror gave place to a strange fascination as she followed Mostyn's plucky and desperate struggle against the two Arabs. She wanted to go to his aid, but her limbs refused the dictates of her brain, apart from the fact that she was without a weapon of any description.

As in a hideous dream she saw the Wireless Officer struggle until he had overcome his antagonists, only to be attacked by the captain of the dhow and the Arab who had returned from his task of securing the lascars.

The period of trance-like inaction passed. Olive stole stealthily towards the three combatants with the desperate intention of throwing herself upon the captain, as he manoeuvred for an opening. She saw the iron bar descend and Peter's automatic slither along the deck. The Arabs, too intent upon settling with the Englishman, paid no attention to the little weapon.

Swiftly the girl grasped the automatic. Even in her haste she remembered to release the safety-catch and to see that there was a cartridge in the breech.

Levelling the pistol she pressed the trigger. The Arab captain threw up his arms and staggered upon the almost exhausted Peter, bearing him to the deck together with the fellow whom he had used as a human shield.

Still at a loss as to the outcome of the fight, Olive waited, finger on trigger, watching the writhing forms almost at her feet. Presently the Arab sailor extricated himself and fumbled for the knife in his sash.

Again the pistol cracked, and the fellow collapsed in a limp heap across the body of the captain of the dhow.

Checking her almost irresistible inclination to ascertain whether Peter was dead or alive, the girl made her way aft, remembering that there were five Arabs and that only four had been accounted for.

A loud, very masculine-like voice, uttering a string of curses that would have done credit to a Thames bargee, greeted Olive's ears. As she stooped to clear the low poop she was just in time to see Mrs. Shallop deliver a clean and beautifully timed punch on the point of the Arab's jaw. The luckless fellow, lifted completely off his feet, crashed heavily against the bulkhead and slithered limply upon the deck.

This much Olive saw by the aid of a horn lantern hanging from the deck-beam. Then, as Mrs. Shallop turned, the girl was also aware that there was a knife sticking into the woman's left shoulder.

Olive offered her assistance. Mrs. Shallop, seemingly aware of the knife for the first time, waved her back.

"Nothing to make a song about," she protested in a gruff voice. "When I want your help I'll ask for it—not before."

And with this ungracious refusal Mrs. Shallop went back into her cabin and shut the door; leaving Olive, feeling considerably bewildered now that the reaction was setting in, standing close to the unconscious Arab.

It was some moments before she pulled herself together sufficiently to go on deck. By this time the dhow had run up into the wind and was gathering sternway with her lateen foresail aback. Olive hardly heeded the fact. Her first care was to ascertain whether any of the three were still living.

Peter looked a ghastly sight, a generous portion of his hair torn out by the roots and blood trickling down his forehead.

A hasty examination showed that he was still alive and apparently without serious injury. Olive washed the stains from his face and rested his head on an improvised pillow. Then she went to the assistance of Preston and Mahmed.

With difficulty she removed the collapsed tent, for in the mêlée the Acting Chief had rolled over upon the folds of the canvas. He too looked a pretty object, for the old wounds on his head had reopened, while in addition he had been stabbed. Olive deftly dressed the injuries and turned to Mahmed.

She did not know what to make of the Indian boy. He was so chipped about that she was unaware whether he was alive or dead.

Olive was still engaged in doing her best to patch Mahmed up when Mrs. Shallop appeared upon the scene. Somehow she had contrived to put a dressing over her wound, although it must have been a difficult task to tie the knot that held the bandage in position.

"Bit of a mess, ain't it?" she remarked. "We'd best clean up a bit. How about heaving those blacks overboard?"

"Are they all dead?" asked the girl.

"Not a bit of it," was the unconcerned reply. "But they soon will be, so overboard with them."

"No," declared Olive firmly. "It's not right—it's murder."

"It would have been murder for us if they hadn't knuckled under," rejoined Mrs. Shallop. "When they come to their senses there'll be more trouble, you mark my words."

Olive glanced in the direction of the Arab captain. Already he was showing signs of returning consciousness.

"What's that hatch under the poop, close to your cabin?" she asked.

"How on earth should I know?" retorted Mrs. Shallop. "It's no odds to me what it is."

The girl went aft, lifted the hatch, and lowered the lantern into the cavernous depths. The place was an after-hold, its for'ard end terminating in a strong transverse bulkhead, while the curved timbers and raking sternpost comprised the remaining walls.

"We'll lower the Arabs down that hatch," declared Olive firmly, when she rejoined her companion. "They'll be safe enough in there."

"No; overboard with them," persisted Mrs. Shallop.

"You'll be tried for murder on the high seas if you do," continued Olive.

The threat caused the woman's blood-thirsty schemes to evaporate.

"All right, then," she conceded grudgingly.

With very little assistance Mrs. Shallop dragged the unresisting forms of the five Arabs aft, after searching them in a very methodical fashion for concealed arms. This done, she passed a rope round each Arab in turn and lowered him into the hold; while at Olive's suggestion a stone jar filled with water was placed in their prison.

"Guess they'll be scared stiff when they come to," was Mrs. Shallop's grim comment, as she closed and secured the hatch. "Where's any food? That job's made me feel quite peckish."

She disappeared into her cabin, while Olive, left to her own resources, began her watch and ward by the side of the still unconscious Wireless Officer.

CHAPTER XXXV
The End of the Voyage

Three days later the dhow was bowling along up the Mozambique Channel with the Madagascar coast showing broad on the starboard beam.

Peter was once more in charge of things. He had made a quick recovery from his hurts, although he still experienced a difficulty in swallowing.

Preston too was making favourable progress. His latest wound was a clean cut. Up to the present there had been no complications, and his amateur nurses had good reason to think that none would be forthcoming.

With Mahmed things were different. Twenty-four hours elapsed before he regained consciousness. He was suffering from at least half a dozen deep knife wounds and several others of a lesser degree of danger. In addition to a serious loss of blood, he was in a high fever.

Peter was greatly concerned over the dangerous state of his trusty servant. He had thought of putting into the nearest port in Madagascar and landing Mahmed for medical treatment, but the boy besought Mostyn Sahib so fervently that he should not be left that Peter decided to carry on.

There was no longer any doubt about the dhow's position. On board, Mostyn had discovered, amongst other articles of navigation, a British-made sextant, and, as soon as the Acting Chief recovered sufficiently Preston had fixed the latitude. The absence of a chronometer mattered little, since the Madagascar coast was visible to starboard.

By the aid of Arab charts it was found that the dhow was now within six hundred miles of Pangawani, the nearest port in the Kilba Protectorate, and, indeed, the nearest territory under British rule. Provided the wind held, the dhow ought to reel off those six hundred miles in from five to six days.

Everything considered, Peter congratulated himself. In a stout, weatherly craft, although on very unconventional lines according to British standards, there was little cause for anxiety on the score of danger. There were ample provisions of sorts, and sufficient fresh water to enable the dhow to carry on without being under the necessity of putting into any port to revictual.

The Arab prisoners gave little trouble. Given food and water and medical stores of their own providing, they accepted the changed conditions with typical Moslem fatalism. Twice a day they were allowed on deck singly, ostentatiously covered by Mostyn with his automatic; and, without the slightest show of opposition, they returned to their place of captivity in the hold directly they were so ordered.

Amongst other articles discovered in the Arab captain's cabin was a leather bag, containing gold and silver coins of an approximate value of £120. This Peter placed in a large trunk, which, in default of lock and key, was secured by driving in several long nails. He told no one of his find, but resolved to hand over the money to the port authorities as soon as the dhow arrived at Pangawani.

After distinguishing herself by knocking out her Arab assailant and making herself useful until Peter was able to resume control, Mrs. Shallop had drifted back into her old style. For hours at a stretch she remained in the cabin assigned to her. When she did appear she indulged in outbursts of complaints against everything in general.

Peter now suffered her in silence. He could afford to do so, knowing that within the next few days he would be relieved both of her company and his responsibility.

On the fifth day following the acquisition of the dhow, the Comoro Islands were sighted on the starboard bow. There were now plenty of craft to be seen, from tramp steamers to dhows. Mostyn let them pass without attempting to communicate. A sort of spirit of independence possessed him. Having gone thus far without outside assistance he was determined to see the business through. Had urgent necessity arisen he would have stopped a large vessel and requested medical attention, but Mahmed was making good progress, and was so emphatic in his desire to remain with his master, that any thwarting of his wishes in that direction would have more than counterbalanced any good that a doctor might have done.

It was not until the morning of the eighth day that land was sighted on the port bow. Once again, after days of adventure, Mostyn was gazing upon the African mainland.

"You'll have to be jolly careful how you approach Pangawani Harbour, old son," cautioned Preston for the twentieth time. "For goodness sake don't put the old hooker on the bar and kipper the show."

"I don't intend to," replied the cautious Peter. "The Arab chart isn't much good. It's on too small a scale. I'll bring up and signal for a pilot, unless there's another vessel making the port. If so, I'll follow her in."

As ill luck would have it the wind dropped about midday, and Mostyn had the mortification of seeing the entrance to Pangawani Harbour at less than five miles away, without being able to gain a hundred yards through the water. At times the dhow was appreciably drifting away from the desired haven. Until close on sunset she was becalmed. Then a stiff off-shore breeze sprang up.

There was no help for it. Throughout the night the dhow was under way close hauled, passing and repassing the entrance without being able to cross the bar. Even after the wind had freed her, Peter would not have risked the intricate entrance in the darkness. So, with the roar of the surf borne to his ears, Peter kept watch during the darkness, until dawn revealed the fact that the dhow was immediately abreast of and less than a mile from the actual fairway.

Yet the harbour was denied him. The sea breeze gave place to another calm, and it was not until the sun was high in the heavens that the customary onshore wind began to make itself felt.

There were other craft making the harbour. Several dhows were in sight, their crews, tired of waiting for the breeze, laboriously sweeping the ponderous craft. Farther away was a gunboat, her white-painted sides looking strangely unfamiliar to people accustomed to the "battleship grey" of warships in home waters.

"She's down from Zanzibar," declared Preston. "She's got a soft job nowadays, but those fellows had a sticky time when I was on the coast. No, I don't think she's coming in here, otherwise we might have had a tow in."

The dhow was now gathering way under the fair breeze. A cable's length astern was another dhow, the crew of which had just relinquished their sweeps and were preparing to hoist sail. Mostyn noticed that the white-robed skipper was intently watching him, and that the curiosity was shared by the rest of the Arab crew.

"P'raps he recognizes the old hooker," he remarked to Olive, who was standing with him on the poop. "He'll be puzzling his brains to know what we're doing on board."

Even as he spoke a distinct splash astern attracted his attention. Stepping aft he was just in time to see a brown figure diving into the water in the wake of another who was swimming a good ten feet beneath the surface.

Then there was another splash and the performance was repeated.

"By Jove!" exclaimed Mostyn. "We've been done. Our prisoners are escaping."

"Have escaped," corrected Olive as five heads, appeared above the surface.

One of the Arabs was swimming strongly, at the same time shouting to his compatriots on the nearest dhow. Two others were making slower progress for the reason that each was encumbered by supporting a disabled man.

Without let or hindrance the escaped prisoners gained the dhow astern and were hauled upon deck. Then, putting her helm down, the succouring craft went about and headed for the open sea.

"They've done us in the eye," declared Peter.

"I'm rather glad," said Olive.

"So am I in a way," agreed Mostyn. "Saved us a lot of trouble, handing 'em over, attending their trial, and all that sort of thing. But it's a bit of a mystery how they managed to break out of the ship."

Leaving the lascar at the helm, Peter went below and examined the hatch of the after-hold. It was intact and secured. Raising it he peered below. The mystery was a mystery no longer. Unknown to him there were two square ports right aft and just above the waterline, which, when in harbour, were used to facilitate stowage of cargo. Seizing their opportunity, the prisoners had kept observation until they saw a friendly dhow within easy distance, and had made their escape through one of the ports.

"And I'm also very glad," continued Peter, "that there's a gunboat within sight, otherwise we might have had to try conclusions with a dozen armed Arabs."

He turned to the second lascar.

"Hoist the pilot flag," he ordered.

The pilot flag—S International—was quickly forthcoming. In the absence of a set of signal flags on board, Olive, under Peter's direction, had made the required flag out of some white linen and a square of blue cloth from the Arab skipper's wardrobe.

The signal was answered with far greater dispatch than at Bulonga, and within half an hour the Pangawani pilot boat was alongside.

"Hello!" was the greeting of the dapper clean-shaven official, as he came over the side and regarded with undisguised astonishment the bedraggled and somewhat battered crew of the dhow. "Hello! You look as if you've been in the wars. Where are you from?"

Before Mostyn could reply Preston broke in:

"Davis, old son!" he exclaimed. "Cut the cackle and get us in. I'm dying for a whisky and soda."

"Great Scott!" ejaculated the pilot in astonishment. "Preston, by the powers! We heard that you were lost in the *West Barbican.*"

"All you hear isn't gospel, my bright youth," rejoined the Acting Chief sententiously, as he took a cigarette from the case offered by the port official. "Hardly expected to see you here, if it comes to that."

"They transferred me from Zanzibar in November last," exclaimed Davis. "It's a move up. Here I'm practically my own boss."

He walked towards the tiller, turned on his heel, and glanced shorewards.

"You can tell your fellows to stow sail," he continued. "We'll tow you in."

"By the by," inquired Peter. "What is the date? We seem to have lost count."

"The eleventh of January," was the reply.

CHAPTER XXXVI
A Round of Surprises

During the rest of the day the picking up of dropped threads was a continual source of astonishment to Peter Mostyn, although it was not the first time that he had been cut off from the outside world.

The dhow was berthed alongside the newly constructed wharf, fronting the modest building which housed the customs and port officials of Pangawani. The two lascars were sent to a native merchant seamen's compound, until they could be shipped back to Bombay in accordance with the terms of their engagement. Mahmed, greatly against his wish, was transferred to a native hospital, on the promise given by Mostyn Sahib that he would be allowed to accompany his master as soon as he was able to do so. Mrs. Shallop, declining offers of hospitality from the wife of a Customs officer, betook herself to a small hotel close to the railway station from which the line, broken only at the as yet unspanned Kilembonga Gorge, starts on its eight-hundred-mile run to the provisional capital of the Kilba Protectorate.

Olive Baird, on the other hand, gratefully accepted Davis's offer to stay with his wife until an opportunity occurred for her to take passage home — the opportunity being determined by Peter's ability to accompany her, and thus carry out his promise.

Dick Preston sturdily declined to go into hospital. Already he had arranged to share rooms with Peter at the Pangawani branch of the Imperial Mercantile Marine Club of which both officers were members.

Before Peter relinquished his command, certain formalities had to be gone through, amongst which was the examination of the vessel by the port officials.

The dhow's cargo was small and comparatively worthless. There were no papers to prove her identity or of where she came.

"What's in that chest, Mr. Mostyn?" inquired the official, pointing to the box containing the money, the lid of which Peter had nailed up. "Coin, eh?

All right, we won't open it yet. I'll wait till we get it ashore, but I'll put a seal on it for our mutual safeguard.'"

In fact he affixed three seals bearing the impression of the arms of the Protectorate of Kilba.

"One more thing," continued the port official. "You'll have to make a declaration before the Head Commissioner. I'll come along with you. We may catch him before dinner."

"Not in these trousers," objected Mostyn, indicating his disreputable garments. "And I must go to the post office."

"Right-o," agreed the official cheerfully. "Nothing like killing three birds with one stone. You and I are about the same build. Let me fit you up. Comyn is my tally."

In a very short time obvious deficiencies in Peter's wardrobe were made good. Then, accompanied by his newly found friend and benefactor, he called in at the post office and dispatched a cablegram to his parents.

The message was characteristic of Mostyn. He did not believe in paying for two words when one would do, especially at the rates charged by the cable company. It was simply: "O.K. Peter".

Having discharged this act of filial duty, Mostyn suffered himself to be led into the presence of the Head Commissioner of the Kilba Protectorate, who happened to be on official duty at Pangawani.

With the Commissioner was the Director of Contracts. Both were under thirty-five years of age—Britons of the forceful and energetic type to which colonial development owes so much.

They were sitting at a large teak table littered with papers and documents. The Director of Contracts was reading a typed cablegram.

"Infernal cheek, Carr," he exclaimed to his colleague. "We've no use for cheap German stuff in the Protectorate. We'll turn it down."

The subject of his righteous wrath was a tender from the Pfieldorf Company offering to supply steelwork "exactly according to the plans and specifications of a contract that has unfortunately failed to be executed", delivering the material at Pangawani within thirty-six days of receipt of telegraphic order, for the sum of £55,000.

"Good!" ejaculated the Commissioner. "Tick the blighters off while you are about it. I'd rather see the Kilembonga Gorge unbridged till the crack of doom than have the place disfigured—yes, dishonoured, if you like—by a Hun-made structure. It was a bad stroke of luck when the Brocklington people's stuff went to the bottom of the sea."

The walls and doors of the official buildings were far from soundproof. Peter, standing with Comyn outside the door, heard the words distinctly. To him they conveyed only one explanation: that in transport from Bulonga to Pangawani the vessel chartered for the conveyance of the steelwork had met with disaster.

Comyn tapped at the door and was bidden to enter.

"I've brought Mr. Mostyn to report to you, sir," he explained. "Mr. Mostyn was in charge of the dhow that landed seven survivors of the *West Barbican* this morning."

"We've just been talking of the *West Barbican*, Mr. Mostyn," said the Commissioner. "We were saying how unfortunate it was that an important consignment for us was lost in the ship. By the by, are you any relation of Captain Mostyn, one of the managing directors of the Brocklington Ironworks Company?"

"He's my father, sir," replied Peter. "I'm afraid, though, that I fail to understand your reference to the loss of the steelwork."

"Hang it, man," interposed the Director of Contracts, "surely you ought to know. You were on the ship when she went down."

"And I know it," agreed Peter grimly. "That she went down, I mean. As for the steelwork, that was landed at Bulonga a day or so before the disaster occurred."

"What?" demanded the Commissioner and Director of Contracts in one breath.

Peter repeated his assertion.

"Glorious news!" exclaimed the Commissioner. "Bless my soul, what possessed them to dump the stuff in a miserable backwater in Portuguese territory?"

"That's for you to say, sir," replied Mostyn. "I took in the wireless message when we were a few hours out from Durban. It came from the Company's agent, and obviously must have emanated from here."

"Obviously fiddlesticks!" interrupted the Director of Contracts. "If it had I would have been responsible for it. Fire away, let's have the whole yarn."

For the best part of an hour Mostyn kept his listeners deeply engrossed. The Commissioner completely forgot that there was a meal waiting for him. Here was an enthralling narrative with an unsolved mystery attached.

"Have you any available funds, Mr. Mostyn," he demanded bluntly, when Peter had brought his story to a close.

"Precious little, sir."

"Then let me make an offer. If you accept you will be rendering a public service and doing your father's firm a thundering good turn. You are in no immediate hurry, I take it, to be sent home?"

Peter thought not.

"Good," continued the Commissioner. "In that case you can act as representative to the Brocklington Ironworks Company, and deliver the goods before the contract date. You've a good sixteen days clear. I'll give you a credit note for a thousand pounds, and you can make your arrangements for chartering a vessel to bring the consignment round from Bulonga. As a matter of fact there's the *Quilboma* lying in harbour at the present time, waiting for cargo. She'd do admirably, and you can get quite reasonable terms. Once the jolly old stuff is planked down on the wharf here your father's firm has carried out its obligation, you know."

It did not take long for Peter to accept the offer. He metaphorically jumped at it.

"Right-o," said the Commissioner, as he dismissed his newly accredited agent of the Brocklington Ironworks Company. "Get a move on. Over you go and the best of luck."

Still feeling considerably mystified, Mostyn left the building. Outside he parted with Comyn, the latter impressing on him that he would be only too pleased to be of assistance to him in any matter and at any time during his stay at Pangawani.

Peter went to the post office a second time. Again he cabled to his father, but with a reckless disregard of the money he was putting into the cable company's exchequer. He did not even wait to put the message into code, but stated that the consignment of steel-work had not been lost in the *West Barbican*, but had been landed at Bulonga. He proposed chartering a tramp and bringing the consignment to Pangawani.

"That'll buck the governor up, I reckon," he soliloquized, as he handed in the cablegram.

His next move was to interview the master of the S.S. *Quilboma*, who, as luck would have it, was also part owner, and being badly in want of a cargo agreed to undertake the run to Bulonga and back at a very reasonable figure.

"When can you get under way?" inquired Peter.

"Tide time to-morrow night," was the reply. "Say about six o'clock."

Peter's peregrinations that day were by no means finished. After being held up and interviewed by the local representative of the *Kilba Protectorate Gazette*, who was also a correspondent to one of the principal London dailies, he found out Olive and told her of his latest plans.

"It won't take much more than a week—perhaps less," he explained. "I don't think that in any case you will be able to find a homeward-bound vessel by that time."

"I won't trouble to do so," declared the girl. "Mr. Davis and his wife are no end of good sorts."

Preston received the news of Peter's venture with considerable envy.

"Wish I were fit enough," he remarked; "I'd come along and help you through with it. Keep your eyes open, old man, and see if you can find out anything about the *West Barbican*. It seems to me that somebody in Bulonga might be able to throw out a good hint as to the cause of the explosion. I may be wrong, but those are my sentiments. When do you sail?"

Peter told him.

"That's unfortunate, my lad," rejoined the Acting Chief. "These people here are giving us a lush-up to-morrow evening. Couldn't wait, I suppose?"

Mostyn shook his head.

"Tide time," he replied briefly.

"Any time between six and nine," added Preston. "Ask the Old Man—he's not your boss, you're employing him—to put it off say till a quarter to nine. Then you'll be able to have most of the fun; Miss Baird and Mrs. Shallop will be there, of course, although I guess neither of us is particularly keen on the old woman's presence."

"She turned up trumps when she tackled the Arab," Peter reminded him.

"All right, get on with it," interposed Preston good-humouredly. "It will be an ordeal for me, watching you fellows enjoying yourselves, an' the doctor's shoved me on to a light diet. He didn't want to let me go, but I'll be there, even if it snows ink."

So back to the harbour Mostyn went to interview the skipper of the *Quilboma* once more.

"'Tain't for me to raise objections," declared the captain, "but it's cutting it mighty fine. Fallin' tide's at nine, d'ye see?"

He tilted back his topee and scratched his head.

"Tell you what," he continued. "I'll take her over the bar at seven o'clock and drop killick outside, if 'tis as calm as it is to-day. Mr. Davis's launch can put you off, and then we'll get under way directly you come aboard. Make it four bells, if you like. There won't be much time lost, seeing as I haven't to smell my way out on a falling tide."

The Old Man's assertion that there would be but little time lost finally dispelled Peter's misgivings. He would have foregone the doubtful pleasure of the lush-up ashore rather than have risked the chance of still further delaying the delivery of the Brocklington Ironworks Company's contract; but now, with these reassurances, Mostyn felt that he could accept the hospitality of the new-found friends without any pinpricks of conscience.

Punctually at the time stated Peter presented himself at the club. Already the Head Commissioner and the port officials were there to welcome their guests.

A little later a rickshaw trundled up to the entrance, and Preston put in an appearance, assisted by a couple of the club servants.

Then, in Peter's eyes at least, a radiant vision arrived, as Olive Baird, simply yet daintily dressed in one of Mrs. Davis's evening frocks, and escorted by her host and hostess, was ushered into the ante-room.

Her introduction to the Head Commissioner took a very considerable time—at least Peter thought so—while others of the Pangawani community flocked up to the girl like flies round a honey-pot.

At length the Head Commissioner suggested that it was time to adjourn to the dining-room.

"We're all here, I take it?" he inquired.

"Mrs. Shallop hasn't arrived yet," replied one of his colleagues, who, although deputed beforehand to take the lady into dinner, was in total ignorance of what she was like or of her rather outstanding mannerisms. "We sent a rickshaw to her hotel an hour ago, sir."

Before the Commissioner could make any remark upon the lady's absence a native servant approached, salaamed, and offered a silver plate upon which was a pencilled note.

"Excuse me a moment," said the Commissioner to his guests.

He pulled aside the bamboo chik that separated the ante-room from the foyer. As he strode out Peter noticed that there was a tall man in a drill uniform standing in front of a couple of native policemen.

Mostyn was not in the least curious. He was aware that the leisure time of a highly-placed official is hardly ever free from interruptions upon matters of state. But he was considerably surprised when a couple of minutes later the Head Commissioner pulled aside the curtain and said:

"Mr. Mostyn, may I speak to you for a few moments?"

Peter went out. The uniformed officer and the two policemen were standing stiffly at attention.

The Commissioner without any preamble plunged into facts.

"This is Inspector Williams of the Kilba Protectorate Police Force," he announced. "He holds a warrant for the arrest of Mrs. Shallop, or, to give her—or, rather, him—his correct name, Benjamin Skeets. He is very badly wanted at home for extensive frauds on the United Trusts Banking Company. His partner in crime, Joseph Shales, whom probably you know under the name of Mr. Shallop, is already in the hands of the Union of South Africa Police. I suppose this is news to you?"

"It is, sir," replied the astonished Mostyn.

"You had no suspicion of the true sex of Mrs. Shallop?"

"None whatever."

"Had he any money when he came ashore?"

"Not to my knowledge, sir."

"Well, the fact remains," rejoined the Head Commissioner drily, "that Mr. Benjamin Skeets has given us the slip; although, we hope, we may possibly lay hands on him before long. He can't get very far away. All right, Williams, carry on. Keep me informed directly you hear anything of a definite nature. Come along, Mostyn; we'll rejoin the others. Not a word about this till after dinner."

CHAPTER XXXVII
How the Steelwork Arrived

There was no doubt about it: Mr. Benjamin Skeets was a very crafty fellow. By adopting the disguise of a woman, and acting up to the part of a vulgar parvenue, he had completely covered his tracks, and had thrown dust into the eyes of everyone with whom he had come in contact—up to a certain point and then only with one exception.

Messrs. Skeets and Shale were no mere novices in crime, and their daring *coup* of defrauding the United Trusts Banking Company of the round sum of £30,000, and their subsequent disappearance, had both mystified and astonished the British public by its audacity, and had completely baffled the greatest detective experts of Scotland Yard.

Skeets had lived up to his disguise very thoroughly. Even the subsequent engagement of Miss Olive Baird had been undertaken solely with the idea of elaborating the smaller but by no means unnecessary details of his disguise. Since there was no reliable description of Mr. Joseph Shales, who was the unseen partner in the deal with the banking firm, it was a fairly simple matter for him to get out of the country under the guise of the husband of "Mrs. Shallop".

It had been the intention of the precious pair to leave the *West Barbican* at Cape Town; hence Mrs. Shallop's anxiety to get a wireless message through as soon as the ship came within radio range of Table Bay. But the absence of a reply from Skeets's confederate at Cape Town had so startled the fugitives that they decided to go on until they found a convenient port, preferably in India, where they could lie low and live on their ill-gotten plunder.

The foundering of the *West Barbican* had upset their calculations. Practically the whole of the pair's booty went down with the ship. Mr. Shallop, otherwise Shales, having no further use for his destitute partner, went off in one of the ship's boats which was eventually picked up. Arriving at Cape Town he took the ill-advised step of looking-up a pal. The latter was already languishing in a South African penal establishment, and Mr. Shales, upon making inquiries, was enlightened by an acquaintance of the convict, who chanced to be an astute detective.

The outcome of this meeting was that Mr. Shallop, under the mellow influence of strong waters, said more than he would have done had he been in his sober senses. Recovering from his maudlin state he found himself in custody.

Having no belief in the worn proverb concerning honour amongst thieves, and perhaps fully convinced that his partner in crime had been lost in the disaster to the *West Barbican*, Joseph Shales confessed to a minor part in the United Trusts Bank frauds, at the same time laying the blame upon the missing Benjamin Skeets.

The immediate result was that directly the news was cabled that more survivors from the *West Barbican*, including Mrs. Shallop, had been landed at Pangawani, the Kilba Protectorate Police were instructed to arrest the much-wanted Benjamin.

Before Mostyn left to go on board the *Quilboma* he had an opportunity of saying farewell to Olive, and at the same time telling her of the astounding news.

"And to think that she—or, rather, he—bluffed the whole jolly lot of us," he added. "Even the Old Man and Doctor Selwyn were taken in completely."

"Not all of us, Peter," rejoined the girl softly. "I knew—but not at first."

"By Jove!" ejaculated the astonished Mostyn. "You did? When did you?"

"Not until the *West Barbican* was sinking," replied Olive. "I found it out then: I couldn't help it. Of course, I didn't know exactly what to do, and I knew nothing whatever of the crime she—I mean, he—had committed. But I meant to tell you some day, Peter."

"We are well rid of him," remarked Mostyn.

"Yes," agreed the girl thoughtfully. Then, after a pause, she added frankly. "But if it had not been for Mrs. Shallop I might never have met you, Peter."

Mostyn departed radiantly upon the voyage on which depended the fate of the Brocklington Ironworks Company's contract.

It was not until the day following that Davis, in his official capacity, completed the inspection of the dhow. When he came to knock off the lid of the box in which Mostyn had nailed up the gold and silver coins, he found that, although the seals were intact, the money had vanished.

Davis gave a low whistle.

"That stuff's been lifted before the dhow put into Pangawani," he declared to his assistant. "The seals being intact proves that."

His companion laughed.

"After sneaking £30,000 friend Skeets wouldn't scruple to lift that little lot," he remarked.

"S'pose so," conceded Davis. "We'll go and report the loss; but I'm afraid that Mrs. Shallop has got well away with it this time."

Which was exactly what had happened. As far as the authorities at Pangawani were concerned Benjamin Skeets had vanished, seemingly into thin air. Although the daily train from Pangawani up-country had been rigorously searched at every intermediate station, soon after the flight of the much wanted man was made known, no one unable to give a good account of himself or herself had been discovered. With the exception of the *Quilboma* no vessel had left the port during the previous twenty-four hours. Native police and trackers had scoured the bush for miles in the vicinity of Pangawani without picking up any traces of the fugitive.

*****

Meanwhile Peter Mostyn was speeding south on board the S.S. *Quilboma*. From the moment the harbour launch had placed him on the deck of the tramp outside Pangawani bar, he was entirely cut off from news of the rest of the world. The *Quilboma* was not fitted with wireless, her owners, since the relaxation of Board of Trade regulations on the termination of the war, having dispensed with what they considered to be an unprofitable, expensive, and unnecessary outfit.

The tramp was only of 1500 tons gross register, and with a speed of nine knots. Her engines were of an antiquated, reciprocating type, while her coal consumption was out of all proportion to her carrying capacity. Had she been plying in home waters she would never have passed the official re-survey; consequently her owners, one of whom was her skipper, took good care to confine the *Quilboma's* activities to the Red Sea and Indian Ocean.

In fine weather, and aided by the current constantly setting southward through the Mozambique Channel, the *Quilboma* was actually making between eleven and a half and twelve knots "over the ground". Three days after leaving Pangawani she arrived at the entrance to Bulonga Harbour.

Six hours elapsed before she was berthed alongside the rotting wharf, to dry-out in a bed of noxious mud as the tide left her.

Mostyn got to work promptly, and with his accustomed enthusiasm. He had the good luck to find the Portuguese agent on the spot. The preposterous storage charges were discussed, haggled over, and settled; gangs of native workmen were hired, and the task of loading up the *Quilboma* with her bulky but precious cargo began.

It was now that Peter met with a sudden and unexpected check, for, on inspecting the metalwork, he found that even in a comparatively short time the moist, tropical atmosphere had attacked the steel in spite of the coating of oxide it had received before leaving England.

To deliver it in this state meant a possible, nay, probable rejection by the consignees; but fortunately the skipper of the *Quilboma* rose to the occasion.

"I've a couple o' kegs of oxide aboard," he announced. "Put the niggers on to it, and let 'em give the stuff another coat."

"Over the rust?" queried the conscientious Peter,

The Old Man winked solemnly.

"Who's to know?" he asked. "Paint's like charity: covers a multitude of defects."

"That won't do for me," declared Peter. "I'll have every bit of the scale chipped off before the least flick of paint is put on."

The skipper shrugged his shoulders but refrained from audible comment. Although in his mind he considered his charterer to be a silly young owl, especially as he was bound to a time limit, he had to confess that Mostyn was doing the right thing.

It took the native workmen two days of unremitting toil (Peter and the Portuguese agent took care that it was unremitting) to clean the steelwork and recoat it with oxide. Then the loading commenced.

With the perspiration pouring down his face, Mostyn supervised the removal of the ponderous girders from the enclosure, the Chief Mate being responsible for the storage of the material in the hold.

Presently the Old Man, puffing like a grampus, hurried up to Mostyn.

"Those four long bits won't stow," he announced. "Our main hold ain't long enough, not by five feet."

"Will they stow on deck?" asked Mostyn.

"And capsize the old hooker in the first bit o' dirty weather we run into?" rejoined the skipper caustically. "You don't catch me doing that, my dear sir. We'll have to leave 'em behind, and the *Thylied* can pick 'em up. She's about due to leave Port Elizabeth, and ought to be here in a week's time."

"Look here, Skipper," said Peter firmly. "You contracted to bring this consignment from Bulonga to Pangawani. I gave you the dimensions of the longest girders before we came to terms, and you declared to me that you could stow the whole of the consignment. And you'll have to do it."

"It ain't a matter of life an' death," expostulated the Old Man. "I'll make a liberal abatement in the freightage charges and—

"You won't," declared Mostyn firmly. "You won't, because you've got to ship every bit of that steelwork; so get busy."

The skipper of the *Quilboma* was one of those easy-going, obliging sort of fellows who can rarely make up their minds and act unless dominated by a person of strong, individual character. He was inclined to let things drift, and would assuredly choose the line of least resistance regardless of the consequences. As a navigator he was passable; as a seaman he lacked the alertness and decision necessary to shine at his profession. For years he had been in command of the *Quilboma*, and not once in that time had he found himself in a really tight corner. It was luck—pure luck—which might at a very inopportune moment let him down very badly.

"What do you suggest then?" he growled.

"I suggested deck cargo," replied Peter. "You turned it down. I don't question your authority or your wisdom on that point. The rest is up to you."

"A' right," rejoined the Old Man. "You just hang on here and keep these niggers up to scratch. I'll fix it up somehow."

And "fix it up somehow" he did; for when at sundown Mostyn returned to the ship he found that the long, heavy girders *were* stowed. The Old Man had had the bulkhead between the main hold and the boiler-room cut through—it did not require much labour, so worn and rusty were the steel plates of that bulkhead—with the result that one end of each of the troublesome girders was within six inches of the for'ard boiler.

At length the loading-up was completed. Steam was raised in the wheezy boilers; the Portuguese customs officials were "suitably rewarded", and clearance papers obtained; and at four in the afternoon the *Quilboma* crossed the bar of Bulonga Harbour, starboarded helm, and shaped a course for Pangawani.

Head winds and an adverse current made a vast difference to the speed of the old tramp. She had taken but three days to run south; five days still found her plugging ahead with Pangawani a good fifty miles off.

The *Quilboma* was now making bad weather of it. Her foredeck was constantly under water, as she pitched and wallowed against the head seas. The glass was falling rapidly. Unless the ship made harbour before the threatened storm broke, it would be impossible to cross the bar until the weather moderated.

The Old Man began to look anxious.

At midday Peter was with the skipper on the bridge when the Chief Engineer approached the Old Man.

"Coal's running low," he reported without any preliminaries.

"How long can you carry on for, Mr. Jackson?" inquired the captain.

"For five hours; less maybe," was the reply. "She's simply mopping up coal on this run. Goodness knows why, 'cause I haven't been pressing her overmuch."

The Old Man nodded. He quite understood. To run the antiquated engines at anything approaching full speed ahead might easily result in the patched-up boilers refusing duty altogether.

"Five hours'll about do," he declared. "Keep her at it, Mr. Jackson."

The Chief Engineer departed. He was not so sure that he could "keep her at it". Under normal conditions the coal taken on board at Pangawani ought to have been more than enough for the round trip. Unaccountably the consumption was much above the average, with the awkward result that the bunkers were nearly empty.

"Pangawani ain't Barry Roads," remarked the Old Man to his charterer. "There isn't a tug at Pangawani; but I'd bet my bottom dollar that, if we were this distance from Cardiff, there'd be a round dozen o' tugs buzzing round an' clamouring to give us a pluck in. No, laddie, we'll have to do it on our own, and we'll jolly well do it, too!"

"Evidently the Old Man's got a 'do or die' spasm," thought Peter, bearing in mind his previous experience with the weak-willed master of the S.S. *Quilboma*. "Let's hope it will last."

By four in the afternoon the Old Man sang to a different tune. The *Quilboma* was now within ten miles of Pangawani; but so low was the pressure in her steam-gauges that she was making a bare five knots.

"I'll signal the first old hooker we fall in with and get her to give us a tow," he decided.

"Not much chance of sighting a vessel off Pangawani, is there?" asked Mostyn.

"You never know your luck," quoted the Old Man sententiously, as he stared apprehensively at the storm clouds banking up to wind'ard.

A few minutes later the skipper of the S.S. *Quilboma* underwent another change of character.

He blew the whistle of the engine-room voice-tube.

"How goes it, Jackson? Last shovelful out of the bunker? How are you off for oil? Yes, any sort. Fair amount—good. Well, stand by: I'll fix you up."

The threatening storm had completely roused the Old Man to definite, practical action. He surpassed himself, and, incidentally, surprised himself and others into the bargain.

Shouting to some of the hands he ordered them to bring axes and to smash up one of the quarter-boats.

"Don't stand there lookin' into the air," he bawled angrily. "Lay aft and do what you're told. I know what I'm doin'. Carve up that blank boat and pass the dunnage down to the stokehold, and be mighty slick about it."

The men, realizing the object of what had previously seemed to be a wanton act of destruction, set to work with a will. In a very few minutes the quarter-davits on the port side were looking very gaunt and forlorn, while a good five hundredweights of wood soaked in crude oil helped to feed the ravenous furnaces.

Half an hour later another boat shared the fate of the first, while, in addition, the crew collected various inflammable gear and passed it below, where sweating firemen threw the impromptu fuel into the furnaces. Bales of cotton waste soaked in oil were added to leaven the whole lump, until the *Quilboma's* stumpy, salt-rimed funnel threw out volumes of smoke that spread for miles astern like a grimy, evil-smelling pall.

The *Quilboma* was now within sight of her goal. Broad on the port bow could be discerned the long, low beach fringed with a quavering line of milk-white foam and backed by the waving coco-palms and the picturesque bungalows of Kilba's principal port.

"How long will that little lot last you, Mr. Jackson?" inquired the Old Man per voice-tube. "Forty minutes? Ay, I'll see to that."

He pointed to one of the lifeboats. The deck-hands, grasping the significance of this display of dumb-show, threw themselves upon the boat. Axes gleamed and fell with a succession of mingled thuds and crashes. Planks, timbers, knees, breast-hooks, thwarts, masts, and oars—all went below to the still insatiable maw of the devouring element.

The skipper of the *Quilboma* made no attempt to signal for a pilot. For one reason, he knew the dangerous entrance intimately; for another, it was doubtful whether the pilot could come out and board the vessel. Yet another: the ship could not afford to wait, with her steam pressure falling and the storm perilously close.

"Starboard—meet her—at that—steady!"

The skipper, standing beside the two quartermasters at the helm, was about to take his sorely tried craft over the dangerous bar. It required pluck, but there was no option if she were to make port at all. It had to be now or never, for, if the *Quilboma* failed to make the bar, she would either be dashed to pieces on the reef or drift helplessly at the mercy of the gale.

With the wind now broad on the starboard beam the old tramp rolled horribly. Peter, hanging on to the bridge-rail, fancied that every piece of steelwork in the hold had broken adrift. Groaning, thudding, quivering, swept by sheets of blinding spray, the *Quilboma* staggered towards the danger-zone. At one moment her propeller was almost clear of the water; at the next the labouring engines seemed to be pulled up, as the madly racing blades sank deep beneath the surface of the broiling sea.

Now she was in the thick of it. Tossed about like a cork, wallowing like a barrel, the old tramp was almost unmanageable. One of the quartermasters was juggling with the wheel of the steam steering-gear like a man possessed, as he strove to keep the old hooker on her course. To port and starboard the ugly reef was showing its teeth, as the remorseless breakers crashed and receded with a continual roar of thunder.

Suddenly a thud, different from the rest of the hideous noises, shook the ship from stem to stern. For a moment—to Peter the pause seemed interminable—she seemed to hang up. Then, with a sickening, sideways lurch she dragged over the hard sand into the comparatively deep and sheltered waters beyond.

"Done it, by Jove!" exclaimed the Old Man, as he rang down for half-speed ahead. "We're in."

But he was trembling like a person in a fit.

Twenty minutes later the S.S. *Quilboma* berthed alongside the quay. The order to draw fires was a superfluous one. The furnaces had burned themselves out.

CHAPTER XXXVIII
The Completion of the Contract

It was too late to commence unloading that day. Peter, having notified the authorities of the arrival of the consignment, and having arranged for the Government surveyor to inspect the steelwork on the following afternoon, made his way to the Davis's bungalow.

So far all was well. The time-limit fixed for the delivery of the Brocklington Ironworks Company's contract was still forty-eight hours off, and there was no apparent reason why the stipulated conditions should not be complied with.

Olive greeted him warmly. Mr. and Mrs. Davis made him welcome with typical overseas sincerity, and he spent a most enjoyable evening.

At daybreak gangs of natives were set to work to clear the *Quilboma's* hold. By noon the bulk of the steelwork lay upon the quayside. At four in the afternoon the material was examined, tested, and passed by the representative of the Kilba Protectorate Government, and an hour later Peter sent another cablegram to his father:

"Contract completed O.K. Official confirmation follows."

This pleasurable duty performed, Mostyn went to pay Mahmed a visit. He found his boy progressing favourably, his many wounds having healed without any sign of complications.

"We'll soon be able to send you back to India, Mahmed," said Peter.

"Me no want go India, sahib," protested Mahmed. "Me stay all one-time with you. Me good cook, me wash-brush sahib's clothes. Me do eb'rything."

"But I'm going back to England," announced his master. "There I don't know what will happen. I may not get another ship for a very long time."

"No matter," rejoined Mahmed, with sublime optimism. "Me stay with sahib. Me make *char* for sahib."

Peter left it at that. He little knew that Mahmed spoke with the tongue of prophecy.

Later on in the evening the Head Commissioner sent for him.

"Are you in a pressing hurry to get home, Mr. Mostyn?" he inquired, after congratulating him upon the successful voyage and happy termination of his trip on the S.S. *Quilboma*.

Peter thought not. Providing that he was not detained to give evidence in the Skeets case, he was in no immediate hurry. Apart from the pleasure of meeting his parents again, he was not particularly keen upon returning to England.

He was well aware of the state of affairs in the wireless service at home; how hundreds of skilled operators were "on the beach" through no fault of their own, and that the prospect of immediate re-engagement was very remote. Wireless officers were just now much in the same position as Tommy Atkins. While there was a war on, and wireless men were in great demand for sea-service, the various shipping companies were almost falling over each other and themselves in their efforts to secure skilled operators. Now that the war is ancient history, and sea risks are falling to pre-1914 level, the services of wireless officers are no longer in great demand. The slump in shipping has dealt a severe blow to radio-telegraphists.

"Quite so," agreed the Head Commissioner, when Mostyn had stated his views. "As a matter of fact we are developing wireless communication in the Protectorate as we find it far cheaper than and quite as efficient as ordinary telegraphy. Setting up telegraph posts for elephants and rhinos to butt into is an expensive game. So I sent for you. I can offer you a really good Government appointment, with free quarters, and splendid prospects of rapid promotion. You're just the type of fellow I want; so what do you say?"

Peter did not reply. He was thinking deeply, struggling with a very complex proposition.

"And six months leave in England on full pay every two years, with free passage out and back," added the Head Commissioner, as an extra inducement—a bait that had often beforetimes turned the scale.

"Thanks awfully, sir," said Peter, "but I'd like to have some time to think things over."

"Certainly," agreed the official, but at the same time he felt rather disappointed. He had been fully prepared to find that Mostyn would jump at the tempting offer. According to what he had heard, Mostyn was a man of action. It rather puzzled him that the Wireless Officer should hesitate to close with the offer of a rattling good post. "Take a day to think things over and then let me know."

As soon as the interview was at an end Peter hurried round to consult his older and, perhaps, more experienced chum Preston.

He found the Acting Chief sitting in a deck-chair under the veranda of the club-house. Preston, like Mahmed, was making a rapid recovery, and already he was able to walk for a few yards with the aid of a stick.

"You silly young blighter!" he exclaimed, when Peter told him of his interview with the Head Commissioner. "Why on earth didn't you jump at it? The pay they're offering you is equal to a cool £800 a year at home, to say nothing of extras chucked in. By Jove! If it had been me— — I suppose there aren't any more plums knocking around for a has-been shellback of forty like me?"

"I didn't jump at it, old man," replied Peter slowly. "I couldn't."

"Why not?"

"On Miss Baird's account," explained Mostyn. "You know I promised to see her safely back to England, and I simply couldn't go back on my word."

Preston grunted.

"Is she so very keen on going?" he demanded. "From what I've heard and seen I don't think she is. Look here, Mostyn, old son. I'm going to be the Grand Inquisitor for once, being almost old enough to be your father. Are you fond of the girl?"

"Yes," replied Peter without hesitation. He was sure on *that* point.

"And is she fond of you?" continued the Grand Inquisitor.

"Think so," was the non-committal reply. "Not so sure about it, though," he added.

"I think I am," rejoined Preston, with a dry chuckle. "I've been keeping my eye upon the pair of you for some considerable time back. Look here, old son; you're a decent sort of fellow with a clean run an' all that. That's what counts with a girl, after all's said and done. You've been offered a rattling good berth with nothing of the 'blind alley' touch about it. All you want now is a sheet-anchor—a jolly sensible girl as a life-partner; one with whom you're not likely to part brass-rags in less than a twelvemonth. Bit of a mixed metaphor, isn't it; but you know what I mean? That girl is Miss Baird; so don't stand hanging on to the slack. Ask her to be your wife."

Peter said nothing. He was very agreeably surprised to hear the hitherto matter-of-fact Acting Chief launching out upon such a subject.

"For goodness sake don't think that I'm starting a matrimonial agency stunt, old thing," continued Preston. "I know many a young fellow who's run aground on the rocks 'cause he's been a fool to get spliced without looking ahead. You're different. There, I've had my say. Full speed ahead and you'll win. And good luck to you."

Thanking his old chum, Mostyn went off feeling considerably elated. Preston's views completely coincided with his own, and the Acting Chief's words of encouragement helped to fill up the gap in Peter's resolution.

The ordeal in front of him was a trying one, he expected; far more stupendous and momentous than he had ever experienced. His adventures while on the books of the S.S. *Donibristle* and the S.S. *West Barbican* were light by comparison.

"No use putting things off," he decided; and, acting upon this resolution, he presented himself at the Davis's bungalow.

Not the shadow of a chance did he have to broach the momentous subject to Olive. Davis and his wife were so hospitable that they never left Peter and Olive alone for one moment.

At eleven, with his mind still unburdened, Mostyn returned to his quarters.

At dawn, after a restless night, he arose, bathed, shaved, and dressed, and went out.

He was by no means the only early riser. The white population of Pangawani make a point of getting exercise before the heat of the tropical day. Watching from afar Peter saw signs of activity at the Davis's bungalow. Native grooms were leading three ponies round to the front of the veranda.

Five minutes later Peter strolled, outwardly unconcerned, past the house, just as Olive and her host and hostess were coming out.

"Hello, old man!" exclaimed Davis. "Topping morning, isn't it? We're off for a canter through the orange groves. Come along."

"Yes, do," added the two ladies.

"Delighted," replied Peter.

Davis shouted to a native groom to saddle another pony.

Mostyn eyed the mount with a certain degree of misgiving. He would have been perfectly at home in the saddle of a motor-bicycle at anything up to fifty miles an hour. There the control was entirely in his own hands. A pony, he reflected, isn't a machine; it is an animal possessing brains and possibly an obstinate will. If the brute took it into his head to exceed ten miles an hour Peter wouldn't guarantee to keep his seat. He didn't profess to be a horseman, but in the circumstances he simply had to risk it and take his chance.

His horsemanship was far better than he had expected it to be, although Olive gave him points on the management of a pony. It was an exhilarating canter along the stretch of broad, white sands, followed by a steady climb to the summit of Mohollo Head.

"Pull up for a minute, Olive," suggested Peter. "My pony is a bit winded, I think. Let's admire the view."

Quite naturally the girl fell in with the suggestion. Davis and his wife were still riding on ahead.

It was an ideal morning. The sun was still low in the eastern sky. A fresh breeze stirred the broad leaves of the coco-palms. The foam lashed itself upon the distant reef, while within the rocky barrier the water was as calm as a mill-pond.

"Isn't this topping!" exclaimed Peter, with a comprehensive sweep of his arm.

"Delightful," agreed Olive. "I shall be very sorry to have to say good-bye to Pangawani."

The girl's whole-hearted admiration gave Mostyn the looked-for opening. With sailor-like alertness he seized the opportunity.

"Then why leave Pangawani?" he asked.

Olive looked at him wonderingly.

"What do you mean, Peter?" she asked. "When do you think you will be going home?"

"In two years time, I hope," he replied. "But that depends upon you."

"Upon me?" rejoined the girl, a faint colour stealing across her half-averted face, as she suddenly realized the point of her companion's remarks.

"Well, you see," explained Mostyn, "I've been offered a Government post out here—a jolly good one. I couldn't accept it because I hadn't spoken to you about it. We agreed, I think, that I should be your guardian—'guardian' is a rotten term, isn't it?—until I saw you safely home."

"Don't, please, let that stand in your way," said Olive.

"It will," declared Peter, "unless— —"

*****

Five minutes or so later Davis exclaimed to his wife: "Hello! Where are the others?"

"I don't know," was the reply. "I quite thought they were following. Trot back and see; I'll wait here."

Another five minutes and Davis rejoined his wife. Deliberately he dismounted, charged a pipe, and lit it.

"There's no hurry," he reported. "They're quite all right. I saw from a distance that I was *de trop*, so I beat a strategic retreat."

Davis finished his pipe, filled up and lit another.

At length the sound of the now walking ponies' hoofs upon the soft ground announced the arrival of the laggards. Then into the glade rode Peter and Olive, both looking radiantly happy.

"Congratulate me, old man!" said Peter excitedly He did not need to explain.

Davis rammed his still-burning pipe into his pocket—he had good cause to remember it later—and extended a sun-burnt hand.

"You lucky dog!" he exclaimed.